MAJOR AWARD DECISIONMAKING AT THE NATIONAL SCIENCE FOUNDATION

Panel on NSF Decisionmaking for Major Awards

Committee on Science, Engineering, and Public Policy

National Academy of Sciences
National Academy of Engineering
Institute of Medicine

NATIONAL ACADEMY PRESS
Washington, D.C.   1994

National Academy Press 2101 Constitution Ave., NW Washington, DC 20418

NOTICE: The project that is the subject of this report was approved by the Governing Board of the National Research Council, whose members are drawn from the councils of the National Academy of Sciences, the National Academy of Engineering, and the Institute of Medicine. It is a result of work done by an independent panel appointed by the Committee on Science, Engineering, and Public Policy, which has authorized its release to the public. The members of the panel responsible for this report were chosen for their special competencies and with regard for appropriate balance.

This report has been reviewed by a group other than the authors according to procedures approved by a Report Review Committee and by the Committee on Science, Engineering, and Public Policy. Both consist of members of the National Academy of Sciences, National Academy of Engineering, and Institute of Medicine.

The **National Academy of Sciences** is a private, nonprofit, self-perpetuating society of distinguished scholars engaged in scientific and engineering research, dedicated to the furtherance of science and technology and to their use for the general welfare. Upon the authority of the charter granted to it by the Congress in 1863, the Academy has a mandate that requires it to advise the federal government on scientific and technical matters. Dr. Bruce M. Alberts is president of the National Academy of Sciences.

The **National Academy of Engineering** was established in 1964, under the charter of the National Academy of Sciences, as a parallel organization of outstanding engineers. It is autonomous in its administration and in the selection of its members, sharing with the National Academy of Sciences the responsibility for advising the federal government. The National Academy of Engineering also sponsors engineering programs aimed at meeting national needs, encourages education and research, and recognizes the superior achievements of engineers. Dr. Robert M. White is president of the National Academy of Engineering.

The **Institute of Medicine** was established in 1970 by the National Academy of Sciences to secure the services of eminent members of appropriate professions in the examination of policy matters pertaining to the health of the public. The Institute acts under the responsibility given to the National Academy of Sciences in its congressional charter to be an adviser to the federal government and, upon its own initiative, to identify issues of medical care, research, and education. Dr. Kenneth I. Shine is president of the Institute of Medicine.

The **Committee on Science, Engineering, and Public Policy** is a joint committee of the National Academy of Sciences, the National Academy of Engineering, and the Institute of Medicine. It includes members of the councils of all three bodies.

Sponsor: This study was funded with Federal funds from the National Science Foundation (NSF) under contract number LPA-9123428. The contents of this report do not necessarily reflect the views or policies of the NSF, nor does mention of trade names, commercial products, or organizations imply endorsement by the U.S. Government.

Library of Congress Catalog Card Number 94-66065
International Standard Book Number 0-309-05029-4

Copyright 1994 by the National Academy of Sciences. All rights reserved.

Available from: National Academy Press, 2101 Constitution Avenue, N.W., Washington, D.C. 20418

B-274

Printed in the United States of America

PANEL ON NSF DECISIONMAKING FOR MAJOR AWARDS

ROBERT H. RUTFORD (*Chair*), President, University of Texas at Dallas, and Chairman of the Polar Research Board, National Research Council

CLARENCE R. ALLEN, Professor of Geology and Geophysics, Emeritus, California Institute of Technology

ALBERT A. BARBER, Special Assistant to the Chancellor, University of California, Los Angeles—Washington D.C. Office

HARVEY BROOKS, Gordon McKay Professor of Applied Physics, Emeritus, in the Division of Applied Sciences, and Benjamin Peirce Professor of Technology and Public Policy, Emeritus, John F. Kennedy School of Government, Harvard University

CHRISTOPHER COBURN, Director, Public Technology Programs, Battelle Memorial Institute

SUSAN E. COZZENS, Associate Professor, Science and Technology Studies Department, Rensselaer Polytechnic Institute

FRANK D. DRAKE, Professor of Astronomy and Astrophysics, University of California, Santa Cruz, and President, SETI Institute

DONALD S. FREDRICKSON, President, D.S. Fredrickson Associates, Inc.

FREDRICK S. HUMPHRIES, President, Florida A&M University

ANITA K. JONES, Chair, Department of Computer Science, University of Virginia (resigned May 31, 1993, to become Director of Defense Research and Engineering, Department of Defense)

LARRY K. MONTEITH, Chancellor, North Carolina State University

DOUGLAS D. OSHEROFF, Professor, Department of Physics, Stanford University

JUDITH A. RAMALEY, President, Portland State University

LYLE H. SCHWARTZ, Director, Materials Science and Engineering Laboratory, National Institute of Standards and Technology

Staff

MICHAEL McGEARY, Study Director
ELIZABETH BLOUNT, Senior Project Assistant

COMMITTEE ON SCIENCE, ENGINEERING, AND PUBLIC POLICY

PHILLIP A. GRIFFITHS (*Chair*), Director, Institute for Advanced Study
ROBERT McCORMICK ADAMS, Secretary, Smithsonian Institution
BRUCE M. ALBERTS, President, National Academy of Sciences *(Ex-Officio)*
ELKAN BLOUT, Harkness Professor, Department of Biological Chemistry and Molecular Pharmacology, Harvard Medical School
FELIX BROWDER, Department of Mathematics, Rutgers University
ROBERT A. BURT, Alexander M. Bickel Professor of Law, Yale Law School
DAVID R. CHALLONER, M.D., Vice President of Health Affairs, University of Florida
ALBERT M. CLOGSTON,* Member, Center for Material Sciences, Los Alamos National Laboratory
F. ALBERT COTTON, Distinguished Professor of Chemistry; Director, Laboratory for Molecular Structure and Bonding, Texas A&M University
ALEXANDER H. FLAX, Senior Fellow, National Academy of Engineering
RALPH E. GOMORY, President, Alfred P. Sloan Foundation
THOMAS D. LARSON, Consultant
JOHN L. McLUCAS,* Aerospace Consultant
MARY JANE OSBORN, Department of Microbiology, University of Connecticut Health Center
C. KUMAR N. PATEL, Vice Chancellor, Research Programs, University of California, Los Angeles
PHILLIP A. SHARP, Head, Department of Biology, Center for Cancer Research, Massachusetts Institute of Technology
KENNETH I. SHINE, President, Institute of Medicine *(Ex-Officio)*
ROBERT M. SOLOW, Institute Professor, Department of Economics, Massachusetts Institute of Technology
H. GUYFORD STEVER, Member, Carnegie Commission on Science and Technology
MORRIS TANENBAUM, Vice President, National Academy of Engineering

ROBERT M. WHITE, President, National Academy of Engineering
  (Ex-Officio)
SHEILA E. WIDNALL,* Associate Provost and Abby Rockefeller Mauze Professor of Aeronautics and Astronautics, Massachusetts Institute of Technology

Staff

LAWRENCE E. McCRAY, Executive Director
BARBARA A. CANDLAND, Administrative Coordinator

* Term expired June 30, 1993

Preface

Under the guidance of the National Science Board (NSB), the National Science Foundation (NSF) supports science and engineering research and education projects. NSF does not carry out these projects itself. It chooses the best proposals submitted by researchers in universities, colleges, and other research institutions. NSF uses a merit review process to identify the most promising projects to receive funding awards. Merit review has two distinctive features: it relies on independent outside peer reviewers to assess the quality of proposals, and it uses criteria that emphasize technical quality and also promote other goals of the nation's research base such as equal opportunity, human resource development, and a broader geographic and institutional infrastructure.

Most of the awards made by NSF are to individuals or to small groups of scientists and engineers. This report addresses a small but important set of awards—very large awards for major research facilities, interdisciplinary research centers, and other large-scale research-related activities. Because of their budgetary impact and importance, it is critical that these major projects be carefully chosen on the basis of their contributions to the nation's research enterprise and not according to political, bureaucratic, or other considerations. To achieve this, major award proposals are subjected to a merit review process. Merit review of major awards is more complicated and sometimes more controversial than that for individual investigator and small group awards.

This report is based on an 18-month study of the NSF-NSB system for making major awards. The study was undertaken by a broad-based expert group, which makes a series of recommendations for improving the planning, selection, and renewal of such awards. The recommendations appear in the chapters on these topics and are summarized in the executive summary.

The panel would like to thank the individuals who took the time to meet with us and share their knowledge, experiences, and views. Special thanks go to Alan M. Gaines, assistant for science and technology to the director of NSF, and NSF liaison official for this study, who made sure we had full and timely access to the publicly available information needed for the study.

The panel was briefed on the decisionmaking process for major awards at its first meeting by NSF and NSB officials: Frederick M. Bernthal, deputy director (chair, Director's Action Review Board); William C. Harris, assistant director for mathematical and physical sciences; Mary E. Clutter, assistant director for biological sciences; Joseph Kull, chief financial officer (executive secretary, NSB Committee on Programs and Plans); and Marta Cehelsky, NSB executive officer. Then-director Walter E. Massey met with the panel at a later meeting. Warren J. Baker, chair, NSB Committee on Programs and Plans, also briefed the panel on how the NSB reviews major award proposals. Former NSF director John B. Slaughter, who recently chaired the site selection committee for the NSF-supported Laser Interferometer Gravitational Wave Observatory, graciously provided his views on the evolution of the major award review process in an interview with a panel member.

The staff would also like to thank others at NSF who provided information and insight: Robert P. Abel, Charles N. Brownstein, Thomas N. Cooley, Peter W. House, Madeleine E. Hymowitz, James M. McCullough, Lynn Preston, and Joanna E. Rom. Susan E. Fannoney of the NSB staff was especially helpful in locating and providing NSB documents relating to NSB review and approval of 10 case study awards; Florence Heckman, NSF librarian, pointed the way to materials on the history of proposal review at NSF; and George Mazuzan, NSF historian, provided access to the historical files of NSF.

The panel appreciates the efforts of Michael McGeary, the study director, who pulled together a remarkable amount of information on

PREFACE

NSF's merit review policies, procedures, and practices, as well as the 10 case study award decisions, and assisted the panel in drafting the report. Elizabeth Blount, staff associate, took care of the many administrative details of panel meetings and report production with skill, energy, and unfailing good cheer. Jeffrey D. Porro, consultant, edited early drafts of the report. We are also grateful for the support and assistance of the staff of the Committee on Science, Engineering, and Public Policy, including Lawrence E. McCray, executive director, and Barbara Candland, executive assistant, and of Philip M. Smith, executive officer, National Academy of Sciences. Florence Poullin was copy editor. National Academy Press staff who helped turn the report into a book included Stephen Mautner, Dawn Eichenlaub, and James Gormley.

Finally, I would like to thank the panel members for their willingness to devote considerable time to the study and for their contributions to this report. The recommendations reflect their vast experience and wisdom and their desire to give NSF and the NSB constructive advice for better decisionmaking on major awards. Panel member Anita K. Jones was able to participate substantially in the drafting of the report before resigning in May 1993 to become the director of defense research and engineering in the Department of Defense.

Robert H. Rutford
Chairman of the Study Panel

Contents

    EXECUTIVE SUMMARY   1

1 MAJOR AWARDS AT NSF   13
    Overview of Major Awards, 15
    Major Awards and Merit Review, 19
    Major Awards and the NSB, 22
    NSF Organization and Staffing for Merit Review, 24
    Overall Conclusions, 31

2 PLANNING MAJOR PROJECTS   39
    Background: Project Planning and Budgeting at NSF, 39
      Long-Range Planning at NSF, 40
      Annual Budget Process, 43
    Major Project Planning and Budgeting, 44
      Capital Facilities Planning, 49
    Findings and Recommendations on Planning and Budgeting, 51
    Recommendation 1: Justification for Major
      Project Awards, 54
    Recommendation 2: Involvement and Support
      of Research Community in Planning, 57

3 AWARDING MAJOR PROJECTS: CRITERIA AND
   REVIEW PROCEDURES   61
    Background: The Merit Review Process at NSF, 62
    Current Review Criteria, 63
    Review and Selection for Major Project Awards, 66
    Findings and Recommendations on Criteria, 70
    Recommendation 3: Primacy of Technical Merit
      Criteria, 71
    Recommendation 4: Human Resource Development
      and Equal Opportunity as a Criterion, 73

Recommendation 5: Cost Sharing as a Criterion, 74
NSF Procedures for Reviewing Proposals, 76
    Proposal Review Process, 76
    Policies and Procedures for Dealing with Bias
        and Conflict of Interest, 80
    Award Decisionmaking, 80
    Findings and Recommendation on Review Procedures, 83
    Recommendation 6: A Two-Phase Merit Review Process, 84

4 **AWARDING MAJOR PROJECTS: NSB ROLE, REVIEW PROCESS DESIGN, AND DECISION DOCUMENTATION**   89
    NSB Role and Procedures, 89
        Findings and Recommendations on the NSB Role, 93
        Recommendation 7: Reorienting the NSB Workload, 93
    Designing the Review and Solicitation Process, 94
        Proposal Review Planning Requirements, 94
        NSB Approval of Solicitation Announcements, 97
        Findings and Recommendations on Proposal Review
            Planning, 101
        *Recommendation 8: Planning the Review Process
        and Criteria*, 101
    Documenting Award Decisions, 104
        Findings and Recommendations on Award Documentation, 106
        *Recommendation 9: More and Better Public
        Documentation of Award Decisions*, 106

5 **RECOMPETITION OF AWARDS**   109
    Project Continuation at NSF, 109
        Findings and Recommendations, 113
        Recommendation 10: More Recompetitions, 113

6 **LOOKING TO THE FUTURE**   117

APPENDIXES
A   BIOGRAPHICAL SKETCHES OF PANEL MEMBERS   121
B   MAJOR AWARDS SUPPORTED BY NSF   127
C   AWARDS APPROVED BY NATIONAL SCIENCE
      BOARD, FY 1986-1992   137
D   MAJOR AWARD CRITERIA FROM RECENT
      SOLICITATION ANNOUNCEMENTS   150
E   THE TEN CASE STUDIES   155

REFERENCES   156

MAJOR AWARD DECISIONMAKING AT THE NATIONAL SCIENCE FOUNDATION

Executive Summary

This report assesses and makes recommendations to strengthen the merit review system used by the National Science Foundation (NSF) to make major awards to support important research facilities, centers, and other large-scale research-related activities. The purpose of the recommendations is to ensure that the most meritorious projects are chosen for support, that the selection process is fair in practice and perception, and that the results in each case are clearly and publicly explained. In this way, the effectiveness and accountability of the major award process will be increased, and the confidence of the research community, Congress, and the public in the system will be enhanced.

The United States has built the most successful research system in the world. The use of peer review to identify the best ideas for support has been a major ingredient in this success. Peer review-based procedures such as those in use at NSF, the National Institutes of Health, and other federal research agencies remain the best procedures known for ensuring the technical excellence of research projects that receive public support. Today, the nation is facing serious international economic competition, which extends to scientific and engineering research. To maintain our world class research enterprise, we will have to be more careful than ever to choose wisely the projects that receive support. The difference between an excellent proposal and one that is merely above average is critical in this effort. The merit review system must be maintained and strengthened to perform the function of choosing the best research for public support.

BACKGROUND

During the past decade, NSF has established Engineering Research Centers, Supercomputer Centers, Science and Technology Centers, and other large research centers and facilities. A few awards were controversial, and called into question NSF policies and procedures for making large award decisions. Some of those involving the location of one-of-a-kind national facilities have generated the sharpest questions about selection procedures. Decisions by the National Science Board (NSB) and the NSF to devote substantial resources to some new center programs and very expensive facilities have also raised questions about the adequacy of their planning procedures. The congressional conference report on FY 1991 appropriations for NSF requested a National Academy of Sciences (NAS) study of the criteria weighed in making major awards and an assessment of the roles of outside experts and agency staff in the merit review decisionmaking process at NSF. The NAS agreed to undertake the project because of the importance of merit review for making major research awards. The study was assigned to the Committee on Science, Engineering, and Public Policy (COSEPUP), which is chartered by the NAS, the National Academy of Engineering, and the Institute of Medicine to address important questions that cut across all areas of science and engineering.

COSEPUP, with the approval of the president of the NAS, appointed a panel with a broad range of expertise to carry out the study (Appendix A). The panel studied NSF's policies and procedures governing major awards, defined as those awards for research and related activities that are subject to approval by the NSB because of their cost. Members of the panel consulted with past NSF directors, current officials, and NSB members, and examined in detail 10 case studies of major awards for research centers and facilities (listed in Appendix E). The NSB reviews between 30 and 50 decisions a year on major awards, which account for about 30 percent of NSF's Research and Related Activities budget of $2.0 billion in FY 1994 (recent awards are listed in Appendix C).

The panel carefully examined the cycles that each major award goes through. These included the processes leading to the initial decision to announce a major project; the planning and implementation of the merit review process; the decisionmaking leading to the award; and subsequent decisions to renew, recompete, or terminate a project at appropriate intervals. The panel focused on the roles of expert peer reviewers, staff, outside advisory groups, and NSB in the merit review process, and on the public explanation of the process, and its outcomes.

In addition to examining NSF policies and procedures, and the organization and resources it has to carry them out, the panel focused on the role and capacity of NSB in discharging its legal authority for design of the review process and for approving each major award. At each stage, NSB has an opportunity to approve, cancel, or postpone further action.

FINDINGS AND CONCLUSIONS

The panel concluded that merit review has generally served well to ensure fairness, effectiveness, and efficiency in decisionmaking on research projects over the years, but for major awards the system needs some changes to accommodate evolving conditions and special features of costly large-scale, long-term projects. NSF has successfully made many highly visible and important awards with relatively few controversies. The merit review system has been the major reason for the high quality of the activities selected for support, and it has served to discourage the use of inappropriate or parochial considerations in the allocation of NSF's research funding. Merit review is not perfect, but no clearly superior method of selecting research and research-related projects for support has been discovered after many years of experience here and abroad.

Although controversial decisions have been relatively rare, they have revealed problems in NSB and NSF policies and procedures that could be avoided. When such problems occur or are believed to occur, they undermine the confidence in the merit review system of the research community, research institutions that compete or hope to compete for major awards in a fair process, and Congress. So far, the success of the merit review system has helped insulate NSF and NSB decisionmaking on major awards from congressional intervention. If confidence in the system is not maintained, the temptation for research institutions to try to have Congress preempt NSF decisionmaking will increase, and to the extent that legislative involvement replaces merit review with political considerations in project selection, the quality of the nation's research system may be negatively affected.

The panel recommends a number of changes to strengthen or improve the planning, review and selection, and subsequent renewal of major awards. Detailed recommendations are contained in various chapters of the report, but the key points follow:

Clear Rules of the Game

The "rules of the game" (i.e., the criteria, procedures, and roles of participants in the merit review process) must be absolutely clear in advance.

In some cases, the criteria or requirements needed to meet them have not been clear or were seemingly redefined during the review process. Although too much detail in specifying criteria might limit the flexibility to respond to innovative proposals, we concluded that to increase procedural fairness, NSB and NSF should be more precise about the criteria and review process to be used. In particular, the primary technical criteria as distinct from other criteria to be considered in the merit review process should be identified in advance in each case.

The panel recommends stronger planning efforts that would help contribute to clearer criteria (Recommendation 1). The panel also recommends that NSF concentrate more effort in designing a better-understood review process for each major award (Recommendation 8).

Primacy of Technical Merit

Technical merit must be the primary consideration in making awards.

The panel strongly supports the primacy of technical merit in the selection of major projects (Recommendation 3), and it endorses the use of a two-phase review process that would clearly indicate the ranking of projects on technical merit before other merit factors are considered (see next section).

Technical merit must be paramount to maximize the likelihood that the project will achieve its substantive research goals. Other criteria of merit should also be given due consideration in selecting the overall winner or winners, but any project receiving an award should rank among the very highest in technical quality. That should be made clear to all reviewers and decisionmakers, along with a sense of the nature and relative priority of each of the criteria.

NSF and NSB must be clearer in each case about the relative priority of the various criteria used, especially of the technical relative to the nontechnical criteria. Otherwise, the weightings of criteria are implicit and can shift continually at the discretion of individual reviewers and program staff.

Appropriate Roles of Peer Reviewers and Staff

The review process must be structured so that the roles of peer reviewers and staff in evaluating and recommending proposals are clearly understood, and trade-offs among technical and other criteria are clearly explained, at each subsequent level of decision-making.

Currently, the summary rating and ranking of proposals by staff at various decision points does not always distinguish peer review from staff judgments. Although staff should make their best case for a recommended decision, the NSF director and the NSB should always know the results of the peer review.

The two-phase review process, properly documented, would make it easier to implement this objective (Recommendation 6). This two-phase process would facilitate the preparation of a summary document that explains the rationale for the decision, including the treatment of peer review results and the trade-offs made between technical and nontechnical criteria in reaching the final decision (see next section).

Public Documentation of Decisionmaking

There should be a public document explaining the results of the review and the rationale for the final decision by the NSB.

NSB minutes rarely record the basis for a major award decision, and no public document of explanation for the final decision is prepared or disseminated. The lack of such documentation leads to public confusion and controversy that could be avoided.

The panel recommends a short, carefully prepared memorandum that summarizes the results of each stage of the merit review process and outlines the rationale for choosing a winning proposal (Recommendation 9). Such memoranda would increase public understanding of major award decisions and therefore enhance public confidence in the system that produces them.

More Stringent Setting of Priorities

Decisions to solicit proposals for very large major awards should take into account their impact on NSF's overall program as well

EXECUTIVE SUMMARY

as on the particular research field involved, and they should be contingent on the realization of expected funds and technological progress.

Careful front-end planning, combined with broad consultation with affected research communities and constant evaluation of priorities at each decision point, must be a part of the process of soliciting and reviewing proposals for a very large major award. Solicitations for awards that have serious long-range budget implications must be based on a broader range of considerations than in the past. The priority within a given field should be clearly established and compared with the overall priorities of NSF across fields. After initial approval of a large project, contingency plans for possible unfavorable program or budget developments should be made for each project and updated annually. The potential impact on NSF priorities if there are unrealized budgetary expectations or unexpected technological problems or opportunities should be carefully reviewed at each decision point. In this way, NSF and NSB would avoid letting a series of small decisions in the development of a major project result in a project that no longer matches the agency's overall program priorities or budget.

The panel calls for stronger planning efforts, including contingency plans for lower funding levels than expected (Recommendation 1), based in part on a broader range of input from research communities affected directly and indirectly by a major project (Recommendation 2). NSB should also put more emphasis on its long-range planning and priority-setting activities (Recommendation 7) and should periodically reconsider the contribution of every project to agency priorities as part of a more systematic project renewal process (Recommendation 10).

The panel understands that its recommendations cannot guarantee a perfect result or prevent individuals and institutions who are denied awards from complaining about the system. This is especially true of awards for large, one-of-a-kind national facilities that must satisfy many expectations. We believe that the changes recommended in this report will result in a fairer and more understandable process and will increase confidence in, and support by, fair-minded participants and interested groups.

RECOMMENDATIONS

Recommendation 1: Justification for Major Project Awards

The NSB should ensure that the large-scale research-related projects that result in major awards are well justified and planned—that is, each is (a) scientifically justified, (b) technically feasible, (c) designed to enhance other activities already in place to achieve the proposed project's goals, (d) of high national priority, and (e) the subject of careful contingency planning.

Recommendation 2: Involvement and Support of the Research Community in Planning

The NSB and NSF should make stronger efforts to see that the basis for initiating large-scale activities is well explained, understood, and accepted to the extent possible by affected research communities. NSB and NSF should take steps to ensure broader consultation with relevant communities beyond those benefiting directly from a major project award, including educational, governmental, and industrial organizations and institutions.

Recommendation 3: Primacy of Technical Merit Criteria

The NSB and NSF should continue to make technical excellence the primary criterion in evaluating the merit of proposals for major awards. To ensure that research funding is used most effectively, no major award should ever be made to a project that

EXECUTIVE SUMMARY

is not of very high technical merit. Additional criteria should be used only to choose the best overall proposal from among those whose technical merit is among the most highly rated.

Recommendation 4: Human Resource Development and Equal Opportunity as a Criterion

The contribution of every major award proposal to overall human resource development should be emphasized. The number of students to be involved—and the inclusion of minorities and women at all levels, from students to senior investigators and project managers—are important components of human resource development and equal opportunity. They should receive more explicit attention in the review process.

Recommendation 5: Cost Sharing as a Criterion

Cost sharing should be used only to demonstrate commitment to the project's goals and never simply to extend NSF funds. Where cost sharing is required, NSF should spell out its expectations in the solicitation announcement. The amount of credit for cost sharing for purposes of evaluating proposals should be stated clearly and capped at a reasonable level. Due account should be taken of the likelihood that cost-sharing commitments will in fact be met in the out years.

Recommendation 6: A Two-Phase Merit Review Process

For major awards, the peer review part of the merit review process should be conducted in two phases. The first phase would be a strictly technical review; to help assure the primacy of technical merit, only those proposals judged to be technically superior would be forwarded to the second phase for any further consideration. In the second phase, the additional merit criteria would be weighed and balanced with the technical criteria by a more broadly constituted group of reviewers. This second-phase panel would recommend the proposal (or proposals) best meeting the full set of criteria. If the proposal judged to have the highest merit overall is not the one ranked highest in the first phase of review for technical merit, the second-phase panel must explain its recommendation fully. If the top-ranked technical proposal is subsequently not recommended by NSF staff, the chair of the first-phase panel or another member of that panel should present the case for it at the NSB level.

Recommendation 7: Reorienting the NSB Workload

NSB should manage its proposal review workload to ensure that adequate time is left for its most important activities of broad policy direction, long-range planning, and program oversight. That could be accomplished by using its exemption authority more frequently, by raising the delegation threshold, or both.

Recommendation 8: Planning the Review Process and Criteria

NSF and NSB should further strengthen their effort to implement a review process for each major award that (a) imposes a reasonable schedule, (b) identifies the appropriate selection criteria and their relative priority, (c) uses the two-phase review process, (d) selects appropriate reviewers to address each criterion at each stage, and (e) is assisted by a central office for review of major projects that ensures quality and consistency based on extensive experience with such complex project reviews.

Recommendation 9: More and Better Public Documentation of Award Decisions

The review and award process should be fully documented and the results made more accessible than is standard or necessary for traditional individual investigator proposals. This process includes such documentation as site visit and panel reports, and the staff-prepared director's memorandum to the NSB summarizing the review results and recommending the awards. In particular, as recommended above, any decision to pass over the proposal rated highest technically (Phase 1) or to recommend a proposal other than the one selected in Phase 2 of the merit review process must be fully explained, and relevant documents should be publicly available.

Recommendation 10: More Recompetitions

The initial planning of every major award should specify the conditions for renewing, recompeting, or terminating the project. As a general rule, each project (or perhaps, in the case of large national facilities, its management) should be openly recompeted within a time period appropriate to the nature of the activity. Such periodic recompetitions should be preceded by an assessment of whether such an activity, however successful, is still needed or whether the funds would be better used in research areas of higher priority or for other mechanisms (e.g., grants to individual investigators instead of a research center, or a program of university instrumentation awards in place of a central national facility).

1 Major Awards at NSF

This report originated in a request by congressional conferees on appropriations for the National Science Foundation (NSF) that the National Academy of Sciences (NAS) study the criteria weighed by the NSF in making major award decisions and assess the roles in the merit review decisionmaking process of outside scientists and executive agency staff. The study was assigned to the Committee on Science, Engineering, and Public Policy (COSEPUP) of the NAS, which has the task of addressing important questions that cut across all areas of science and engineering, such as peer review of research proposals.

COSEPUP agreed to study NSF's system of decisionmaking for major awards, with the understanding that "major" awards were those subject to review by the National Science Board (NSB) because of size (at least $1.5 million a year or $6 million over five years) and funded from NSF's appropriation for Research and Related Activities (R&RA). The report does not address decisionmaking on awards for projects funded from the appropriations for Education and Human Resources or the U.S. Antarctic Program, although the recommendations may also be adaptable to those projects.

The NSB has and generally uses the authority to review and approve all major awards. Currently NSB reviews between 30 and 50 decisions a year, mostly involving large research centers and research facilities. NSB-approved awards constitute about 30 percent of NSF's R&RA budget of $2.0 billion in FY 1994. Authority for the other 8,000 or so awards made annually, which average less than $90,000 a year, is delegated to the director of NSF (although the awards must be made within program areas that have been approved by the NSB).

COSEPUP also obtained NSF's agreement that the report would not comment on the substantive merits or "correctness" of any particular award decision, although the immediate impetus for the congressional request resulted from a specific case (i.e., the decision to award the National High Magnetic Field Laboratory to a university that was not the first choice of outside peer review panels). Evaluating the merit of even one specific decision would require a full-scale peer review process parallel to the original one, an effort beyond the time and resources available for the study. COSEPUP concluded that it was more useful and appropriate to evaluate NSF's *capacity* to make wise decisions in the future on awards (i.e., to see if its review policies, procedures, structure, and resources are conducive to good decisionmaking). Even the best merit review decisionmaking process for major awards cannot guarantee a perfect result; the projects are too complex and the knowledge is too imperfect for that. Nor will an excellent merit review process always prevent complaints from applicants who are denied a major award, but the basis for the decision should be understandable to fair-minded observers. If the process is fair and understandable, NSF will also remain free of the pork barrel pressures that have affected other science appropriations.

COSEPUP, with the approval of the president of the NAS, appointed a panel to carry out the study. The panel consisted of 14 experts in physics, astronomy, geosciences, engineering, biology, and social science research; science policy and peer review; organization and management of federal research agencies, academic institutions, and large research projects; and federal grant and contract administration. The panel was also constituted to have members from institutions and regions that have received relatively few NSF awards, as well as from those that have received many over the years. An effort was made to appoint members who were not direct participants in programs resulting from major awards. A few members, however, were from institutions involved in major projects supported by NSF; they provided a grantee perspective on NSF award decisionmaking.

The panel studied NSF's policies and procedures governing major awards; consulted with past NSF directors, current staff, and NSB members; and examined 10 case studies of past major awards for research centers and facilities (Appendix E). It studied the series of cycles that a major award undergoes—the initial decision to launch a major project; the planning and implementation of the review process; the decisionmaking leading to the award; and subsequent decisions to renew, recompete, or terminate a project at the end of each award period—focusing on the roles of peer reviewers, staff, outside advisory groups, and the NSB in merit review. The panel also looked at the extent of involvement in, and understanding of, the process by communities outside NSF.

OVERVIEW OF MAJOR AWARDS

The mission of the NSF is to foster the growth of new knowledge through a balanced program of investments in high-quality science and engineering research projects, education and training programs, and related research infrastructure. To help carry out this mission, NSF plans and makes some major awards for national-scale research facilities, multidisciplinary research centers, and large-scale organized research programs.

Projects supported by major awards vary widely in terms of function, size, uniqueness, size of user community, and sponsorship. Functionally, most are either facilities or interdisciplinary research centers.[1] Operationally, they range in size from the threshold level of $1.5 million in annual funding to $50 million a year for the National Center for Atmospheric Research (NCAR). They also

---

[1] Other types of activities include a large-scale longitudinal social and economic survey (the Panel Study of Income Dynamics run by the University of Michigan); some large university-based disciplinary research groups (e.g., nuclear physics research with electrons, photons, and antiprotons at the University of Illinois); and several institutes for mathematics and theoretical physics.

include multiyear construction projects in a range of sizes up to more than $200 million for the Laser Interferometer Gravitational Wave Observatory (LIGO).

Many of the major awards are part of multiaward programs. There are, for example, 25 Science and Technology Research Centers (STCs), 18 Engineering Research Centers (ERCs), and nine Materials Research Laboratories (MRLs) located at universities around the country. There are four NSF-supported Supercomputer Center facilities.

Some major awards are for one-of-a-kind national facilities based at and operated by one university or operated in a separate location by a consortium of universities. Competitions for single university-based facilities have occasioned the most controversy because the stakes are high and there can be only one winner. Examples include the National Nanofabrication Users Facility (Cornell University), Earthquake Engineering Research Center (State University of New York at Buffalo), National High Magnetic Field Laboratory (Florida State University), and the National Astronomy and Ionosphere Center (operated by Cornell at Arecibo, Puerto Rico). The roles of careful planning, clarity about criteria and their weightings, choice of appropriate reviewers, prior understanding of the review process that will be used, and justification of the final decision are especially important in these cases.

Some of the major NSF-supported projects serve a large number of users and are accessible to any researcher with a suitable project. The Supercomputer Centers, for example, serve all fields of science and are readily accessible by NSFNET to researchers across the country. Most facilities, however, are set up to serve a particular field or subfield.[2] The global seismic detector network operated by

---

[2] Many of the facilities have in-house research staff who work part-time on their own research projects that do not go through the same merit review process an outside researcher undergoes to obtain an NSF award for a project using the facility. This opportunity to access unique facilities enables the facilities to recruit high-quality staff who can better assist outside researchers in using the facilities more effectively as well as develop state-of-the-art instrumen-

the Incorporated Research Institutions for Seismology (IRIS) can be used remotely by earth scientists. LIGO is a specialized facility for gravitational physics research.

Some of the largest awards go to consortia of research institutions. The University Consortium for Atmospheric Research (UCAR), which operates the National Center for Atmospheric Research, is a consortium of more than 50 universities active in atmospheric research. The national optical telescopes at Kitt Peak, Arizona, and Cerro Tololo, Chile, are managed by the Associated Universities for Research in Astronomy (AURA). AURA has 17 university members. IRIS, which is constructing and managing the global network of seismometers, has more than 80 member universities and colleges. The Ocean Drilling Program has seven international partners representing 18 nations that are also members of an international scientific organization, the Joint Oceanographic Institutions for Deep Earth Sampling, which provides scientific advice and direction to the program. A number of ERCs and STCs, although based at one university, involve networks of universities.

Most major awards are solicited (i.e., there is an open competition for which proposals are formally requested). Not all are competed, however. Some of the largest and most expensive facilities have been developed jointly with the particular group of research institutions whose researchers would be likely to make the most use of them. NSF invites the group to incorporate and encourages it to submit proposals to build and manage the project. This was the approach taken with NCAR (UCAR) and the Kitt Peak National Observatory (AURA).

IRIS is a more recent example of an unsolicited proposal. The universities involved in research on the earth's crust were asked to submit a proposal to build and operate the global seismic network and

---

tation and useful data sets, but it reduces the time available to outside researchers. The question of appropriate balance between in-house research and outside users to achieve the most overall progress in a field is one that cannot be determined here but must be resolved during the planning and management of each major project.

manage the data. Although invited, it was reviewed as a traditional unsolicited proposal (i.e., no special criteria or procedures were specified, the four basic review criteria were used, and there were no competing proposals). The same approach was also taken with LIGO. The leading university groups involved in gravitational research decided with NSF's encouragement to submit a joint proposal to build and manage the facility. Proposals to provide the two sites for LIGO were, however, formally solicited in an open competition.

These projects also vary in the expectations and conditions for renewing or reopening them to competition. The traditional approach has been an open-ended one in which awards are expected to be renewed as long as the grantee's performance is satisfactory. UCAR has managed NCAR, and AURA has operated Kitt Peak since 1960 under a series of renewed awards that were not openly competed. A similar approach was taken with the Francis Bitter National Magnet Laboratory at the Massachusetts Institute of Technology (MIT), which was inherited from the Air Force in 1970); and the National Nanofabrication Users Facility at Cornell after it was established through a competitive solicitation in 1977. NSF eventually decided to open those awards to competition.

NSF has taken a different approach in the programs of campus-based research centers. The 12 original MRLs that NSF inherited from the Advanced Research Projects Agency in the Department of Defense began to compete with new proposals for renewal funding, and as a result, some lost NSF support. ERCs and STCs receive five-year awards, but undergo a full merit review during the third year of each award and either receive a new five-year award or are phased out at the end of the current award period. After 11 years, a center is supposed to be on its own or must reapply de novo in competition with other new proposals.

MAJOR AWARDS AND MERIT REVIEW

All awards, large and small, are made in response to proposals that are reviewed for merit by outside peer experts, NSF staff, and in some cases, scientific advisory committees or other federal officials with relevant expertise. This merit review system is designed to ensure that appropriate criteria are used to evaluate proposals and identify the proposal with the best promise of achieving the goals of the project. Merit review, when used for major awards, often involves more diverse criteria and a more complicated process (ultimately involving the NSB) than when it is used for the typical small research project grant.

In a given fiscal year, NSF supports about 18,000 research awards. Annually about half are new awards chosen from among about 24,000 proposals (including renewals). The remainder are continuing awards in their second or later year. Most of the 18,000 awards are grants to individual researchers or small research groups. Some are very large awards for research centers, facilities, and other major projects. Although fewer than 100 major awards of $1.5 million or more are made each year, they account for about 30 percent of annual R&RA budget expenditures.[3] Since awards for research centers and facilities tend to be larger and awarded for longer periods than average, they constitute a greater mortgage on future R&RA funding than individual research grants.[4] Major awards constitute an even larger share of the budgets of certain NSF

---

[3] In contrast, the mean annual size of all awards is less than $90,000. In fact, because most are two-year grants to individual university-based researchers working with a few graduate students and postdoctoral fellows, the median annual award size was about $50,000 in FY 1990 (compared with $51,000 in 1988) (NSF Executive Information System).

[4] For example, in June 1991, NSF projected commitments in FY 1992 and beyond of $1.1 billion, of which $478 million was committed to center and facility awards (NSB, 1991a:C-5); thus, these awards accounted for less than 30 percent of current spending but 43 percent of funding committed in the future.

directorates with major infrastructural support, for example, Computer and Information Science and Engineering (48 percent of current funding and 63 percent of commitments) and Geosciences (31 percent of current funding and 53 percent of commitments).

Major awards are also significant because they usually support activities that promise to have a major influence on the conduct of research in their field—facilities such as state-of-the-art telescopes, supercomputers, a global network of seismic detectors, and research centers that promote interdisciplinary or application-oriented research or both. Unlike the traditional small research project grants to individual researchers and small research groups, major awards may also invite geographic competition and "pork barrel" pressures from Congress because of the economic benefits they may bring to state and regional economies. These situations offer NSF an opportunity to leverage cost sharing from private industry and state governments. They may also provide opportunities for international scientific cooperation and cost sharing.

From the beginning, NSF has tried to ensure the quality of the research that it supports by openly inviting researchers to compete for funding through proposals. NSF has also promoted quality through an evaluation process that includes outside peer review of proposals by active researchers best able to apply the selection criteria. Peer review is a key part of the merit review process, because it increases the objectivity and impartiality of proposal evaluation and helps to ensure the quality of the activities supported by NSF. It helps give award decisions credibility with the scientific community, Congress, and the public.

The peer review aspects of merit review maximize the role of technical considerations in making awards and help shield the decisionmaking process against internal agency bureaucratic interests and outside political pressures. This competitive peer review mechanism was later extended to large-scale research facilities and centers even though the award criteria for such projects tend to be more numerous and varied. Despite the greater complexity and higher stakes involved in major award decisions, NSF has successfully made many without controversy. It has also maintained its autonomy to choose projects to support and proposals to fund without political interference.[5] The criteria, procedures, and management practices associated with decisionmaking for major awards must be clear, effective, fair, and accountable.

Some recent decisions by NSB have raised questions about the review process, selection criteria, choice of reviewers, staff discretion, or the NSB's role. For example, one impetus behind this study was NSF's 1990 decision, approved by the NSB, to award a five-year, $60 million grant to build and operate a National High Magnetic Field Laboratory to a consortium headed by Florida State University, even though peer review groups—the site visit panel and divisional advisory committee—had recommended a proposal by MIT. The NSF-NSB decision in 1990 to award $212 million to build and operate LIGO has been criticized because it may squeeze the budgets for other projects if NSF's funding does not grow as planned. That decision also drew criticism because of questions about the engineering feasibility of achieving the levels of sensitivity needed to detect gravitational waves.

Even if there were no controversial award decisions, a review of NSF's capacity to make wise decisions on major awards would be prudent and timely. The number of major awards and their share of the NSF budget have been increasing in response to several trends. Research facilities and instruments are increasing in capability and cost. They, in turn, foster larger-scale and more interdisciplinary research efforts so that areas traditionally dominated by individual investigator research are becoming dependent on expensive facilities, coordinated research programs, or both. The budgetary impact of the growth of research center programs and the increasing costs of research facilities has been offset by the steady real growth in the NSF budget since 1986, but such growth is not guaranteed.

---

[5] NSF is one of the few agencies that has not been subjected to academic pork barreling (e.g., it has not had funding for facilities at particular universities or research institutions earmarked in its congressional appropriations).

MAJOR AWARDS AND THE NSB

The focus of this report is the set of awards for research and related activities subject to NSB approval because of their size (more than $1.5 million a year or more than $6.0 million over the award period, which may not be longer than five years) or importance. In FY 1992, for example, the NSB reviewed 49 awards—32 for research-related activities and 17 for education-related projects (listed in Appendix C). Over the five-year period FY 1988 through FY 1992, NSB reviewed 157 awards—120 for research, 35 for education, and 2 for logistical support in Antarctica (Appendix C). The 120 research awards involved 91 discrete projects (some projects received multiple awards during the five-year period).

The NSB also reviews and approves "project development plans" (for most major facilities) and solicitation announcements (for major awards that have been formally solicited). It approves new programs, which usually consist of individual investigator grants but may also include new programs of research centers. This report addresses the steps that occur prior to the review of proposals (Chapter 2) because they determine the goals of a major award, which in turn should shape the review process and criteria (Chapter 3). The better planned a major project is, and the better planned the review process is, the more likely is the award decision to be understood and supported in the affected communities.

The NSB has at least three opportunities to review new large-scale activities as they move through the major award process (Table 1.1).[6] First, in the case of physical facilities, the NSB approves early in the process a project development plan that establishes the need and technical feasibility of the facility. In the case of research centers, the NSB receives a short document describing the new program in conjunction with a draft of the solicitation notice.

---

[6] Although the NSB does not approve smaller grants individually, it does approve the *programs* through which such grants are made. In October 1989, for example, NSB approved a new Program for Arctic Social Science.

TABLE 1.1: Decision Points for NSB Involvement in Major Awards

Phase	NSB Action
Planning	**Project Development Plan Approval for Research Facility Awards** Since 1979, NSB has required its review and approval of "project development plans" for "big science" projects involving large-scale commitment of funds; capital facilities and major equipment; multiyear duration; and continuing expenditures for maintenance, replacement, operating costs, and research. **General Program Approval for Center and Other Nonfacility Awards** NSF cannot accept proposals unless the program within which the award will be made has been reviewed and approved by NSB.
Design	**Approval of Solicitations for Major Awards** NSB reviews and approves Requests for Proposals and other solicitations in which the resulting awards are expected to require NSB approval. The solicitations document the specific procedures and criteria that will be used to decide on awards. (Solicitations may be issued before NSB approval in exceptional circumstances as long as NSB is advised at its next meeting.)
Award	**Approval of Proposed Major Award Recipients** NSB reviews and approves decisions to make awards of at least $1.5 million a year or $6.0 million over the award period (up to five years). (NSB may grant a waiver for routine or continuing awards that pose no significant problems or policy issues.)

Second, if proposals for the award are solicited, the NSB approves the solicitation notice, which describes in detail the selection criteria and review process for the particular award. Some major awards are considered investigator initiated rather than solicited; in these cases, there is no special notice soliciting proposals. Third, at the end of the process, the NSB reviews and approves all awards that exceed the delegation threshold of $1.5 million a year or $6.0 million during the award period (which may be up to five years). The responsibility for approving major awards is also closely related to NSB's role in working with the director of NSF in long-range planning, setting program priorities, and developing NSF's annual budget request.

NSF ORGANIZATION AND STAFFING FOR MERIT REVIEW

NSF is organized into *divisions* and *programs*. The divisions correspond most closely to the traditional scientific and engineering disciplines (e.g., astronomy, chemistry, earth sciences, molecular and cellular biosciences, computer and computation research). The division in turn are grouped under six research directorates headed by assistant directors of NSF. The divisions in NSF's research directorates are listed in Table 1.2.[7]

The programs are generally current research areas within disciplines, although some of them include facilities, center programs, and other modes of research support. In the Division of Earth Sciences, for example, there are research grant programs in geology and paleontology, geophysics, petrology and geochemistry, tectonics, and hydrology. The major projects are in two other programs: continental dynamics and instrumentation and facilities. The Physics Division has research grant programs in atomic, molecular, and

---

[7] NSF also has a directorate for education and human resources and two administrative offices that are not included in this study.

TABLE 1.2: NSF Research Directorates and Divisions

Biological Sciences Directorate
    Molecular and Cellular Biosciences
    Integrative Biology and Neuroscience
    Environmental Biology
    Biological Instrumentation and Resources
Computer and Information Science and Engineering Directorate
    Computer and Computation Research
    Information, Robotics and Intelligent Systems
    Microelectronic Information Processing Systems
    Advanced Scientific Computing
    Networking and Communications Research and Infrastructure
    Cross-Disciplinary Activities
Engineering Directorate
    Biological and Critical Systems
    Chemical and Thermal Systems
    Design and Manufacturing Systems
    Electrical and Communications Systems
    Engineering Education and Centers
    Industrial Innovation Interface
    Mechanical and Structural Systems
Geosciences Directorate[a]
    Atmospheric Sciences
    Earth Sciences
    Ocean Sciences
Mathematical and Physical Sciences Directorate
    Mathematical Sciences
    Astronomical Sciences
    Physics
    Chemistry
    Materials Research
Social, Behavioral and Economic Sciences Directorate
    Social, Behavioral, and Economic Research
    International Programs
    Science Resources Studies

---

[a] The Office of Polar Programs was recently transferred from the Geosciences Directorate to the Office of the Director.

optical physics; elementary particle physics; theoretical physics; nuclear physics; and gravitational physics; some of these programs also support major awards for facilities and large research groups.

Program directors (sometimes called program officers or program managers) are key decisionmakers on awards because they make the initial decisions on which proposals to fund, for how much, and for how long. The key role of program directors applies to major awards, which ultimately must be approved by the NSB, as well as to the usual small grants, which are approved at the division level. In fact, the importance of program directors begins even before the award stage. Because they are closest to research and researchers, program directors often propose new projects and draft the substance of project development plans and project solicitations that may result in major awards eventually approved by the NSB. Program directors are also responsible for designing the review process to be used (within approved NSF policy guidelines), selecting reviewers, participating in site visits, and staffing meetings of the review panels. Finally, they are responsible for monitoring the performance of previously funded projects as part of the process of deciding whether to continue, expand or scale back, terminate, or recompete them.

These responsibilities mean that program directors must be well informed about research trends in their areas, knowledgeable as to who the most productive researchers are, and equipped with adequate time and resources to coordinate and synthesize the review processes (see Table 1.3). NSF should have enough program directors to carry out NSF responsibilities, including coordination of the merit review of all proposals.

Out of NSF's staff of about 1,200, approximately 250 are program directors in the six research directorates, who have the authority to manage the reviews of some 24,000 proposals a year and to recommend about 8,000 for funding. Program directors have doctorates or equivalent experience in the field in which they work, and have been active researchers in their field. About one-third are "rotators," visiting scientists and engineers from academia spending a one- or two-year tour of duty as program directors at NSF. NSF

TABLE 1.3: Roles and Responsibilities of Program Directors

Proposal Processing and Evaluation

- Designs and implements the proposal review and evaluation process (within NSF policy guidelines and subject to higher-level approval)
- Prepares program announcements and proposal solicitations (approved at higher levels, including the NSB in the case of a major award)
- Selects appropriate individuals to review proposals as individuals or as members of a panel (within NSF policy guidelines and subject to higher-level approval)
- Conducts panel meetings and/or site visits
- Conducts technical reviews and analyses
- Evaluates external reviews
- Negotiates proposal budgets and work plans
- Maintains liaison and coordination with other federal agencies in connection with duplicate proposals or joint funding of proposals
- Negotiates revised proposal budgets
- Conducts final review of proposals and evaluations, and recommends acceptance or declination
- Prepares documentation of review and decision processes
- Informs proposer about results of review and decision processes

Program Management

- Interacts with the Division of Grants and Awards (DGA) in processing and administering NSF awards
- Keeps abreast of trends and developments within his or her scientific field by reading the relevant literature, attending scientific meetings, and having personal discussions with leaders in the field
- Acts as the principal NSF contact point for the research community
- Recommends new or revised policies and plans in scientific, fiscal, and administrative matters to improve program activities and management
- Initiates new program directions by recommending support of new projects and phasing out old projects
- Represents his or her scientific discipline or area of particular competence in internal NSF consideration of priorities and allocation of resources

- Reviews and evaluates reports and publications submitted by awardees
- Makes site visits and consults with awardees
- Initiates revisions of project budgets and project descriptions when necessary, and gives guidance and management oversight to an extent appropriate and authorized by the conditions of the award
- Provides for dissemination of research accomplishments or other results from awards
- Reviews complete awards, including technical reports, summaries, and journal publications

Consultation and Liaison

- Advises prospective awardees and institutional representatives about NSF objectives, policies, and practices or refers them to DGA
- Serves as primary consultant within NSF on technical matters in his or her area of competence
- On request, coordinates and advises on aspects of his or her program that involve other facets of society such as national resources, technological assessment, and social and cultural organization
- Exchanges program information with other agencies and institutions
- Represents NSF at professional meetings and seminars

Administration

- Formulates plans, supervises program staff, and assigns and reviews work
- Prepares reports
- Fulfills internal budget and operating requirements
- Performs staff work for, and participates in, program review and evaluation activities
- Recommends new, and improves existing, procedures for program management procedures

---

NOTE: Many of the functions listed are not the final responsibility of the program director but involve section heads, divisions directors, assistant directors, and other parts of NSF.

has promoted this practice as a way to bring in program directors with current knowledge of pioneering research and to prevent the agency from becoming too bureaucratic or out of touch. On the other hand, rotating staff makes continuity of policies and procedures more difficult, which may cause inconsistent treatment of the kind of longer-term activities that tend to be supported by major awards.

Two trends could have a negative impact on the ability of NSF program directors to manage the merit review process for all awards, including major awards. First, the number of proposals has increased much faster than the number of staff. Second, travel funds required by program staff (for site visits by program directors to oversee the implementation of major awards, and for participation in scientific conferences and other professional activities that keep them abreast of current research developments and familiar with productive researchers) have not increased.

The number of fully reviewed proposals has gone up 70 percent over the past 10 years. NSF's budget grew 76 percent in constant dollars from FY 1983 to FY 1993. The number of staff, however, decreased slightly during that period, from 1,213 to 1,192. As a result, the percentage of the NSF budget devoted to administration has declined from more than 6 in 1983 to less than 4 in 1993.

NSF has coped with the growing workload by adopting new information technologies and simplifying proposal processing procedures. In 1984 an internal staff task force recommended a number of measures to speed up processing (NSF, 1984b). In 1990 a task force on merit review recommended additional measures to simplify forms and streamline procedures (NSF, 1990c). NSF also has moved to reduce the proposal burden on the research community by reducing the number of pages allowed in a proposal and increasing the average award length to three years (NSF, 1992d).

Despite these steps, NSB member James Powell (representing the NSB chairman) brought up the imbalance between funding and staffing levels as the one concern he could "single out from all the others that the Science Board has" at hearings on NSF's FY 1993 appropriation (U.S. Congress, 1992:5). Powell said the staff was overworked because of the large increase in the number of proposals reviewed and funded. He also brought up the greater extent of active oversight called for by ERCs and other new types of complex activities undertaken by NSF in recent years. At the same hearing, NSF director Walter Massey testified that NSF efforts to cope with growth by investing in computers, electronic mail, and other new technologies had reached their limit, and he asked for an increase in staffing (U.S. Congress, 1992:6).

In addition to automation, NSF has coped with the large increase in number of proposals by increasing the number of professionals within its static staff size. Nevertheless, the average number of proposals per program director has been increasing. Program directors for major awards may handle only a few awards. The large national facilities for astronomy and atmospheric research, for example, which come up for noncompetitive renewal every five years, have several program directors managing their awards. Other program directors handle a small portfolio of center awards that go through a full proposal review every three years. For example, ERCs and Industry/University Cooperative Research Centers are administered by a separate Division of Engineering Education and Centers in the Directorate for Engineering. Supercomputer Centers are under a separate division of the Directorate for Computer and Information Science and Engineering. Some program directors have a mixed portfolio: they oversee a large number of individual investigator proposals and one or more centers or facilities.

Up to now, concerns about excessive workload have focused on programs with large numbers of individual investigator proposals. Nevertheless, if NSF increases its efforts to plan major awards, adopts more elaborate review procedures, or expects greater oversight of ongoing awards as part of the renewal process, the panel believes that the number of staff needed to administer merit review for these awards should be increased.

OVERALL CONCLUSIONS

The panel concluded that merit review has generally served well to ensure fairness, effectiveness, and efficiency in decisionmaking on research projects over the years, but for major awards the system needs some adjustments to accommodate evolving conditions and special features of costly large-scale, long-term projects. NSF has successfully made many highly visible and important awards with relatively few controversies. The merit review system has been the major reason for the high quality of the activities selected for support, and it has served to discourage the use of inappropriate or parochial considerations in the allocation of NSF's research funding. Merit review is not perfect, but no clearly superior method of selecting research and research-related projects for support has been discovered after many years of experience here and abroad.

The panel observed that merit review procedures and criteria governing major awards have been adapted from those originally used for awards to individual investigators. The scientific community perceives the criteria and procedures for merit review as difficult to understand and sometimes inconsistent when criteria in addition to strict technical merit are employed, which is typically the case in large-scale projects. The panel encountered cases in which NSF's objectives and criteria—or their relative importance—seemed to be applied inconsistently during the review process: the combined expertise of the reviewers did not match the ostensible criteria; the relative roles of staff, reviewers, and advisory groups were unclear; and the rationale for the final decision was not clearly stated. To sustain confidence in NSF's review process, such problems need to be systematically reduced for large awards, both in fact and in the perception of the research community, Congress, and other interested groups.

NSF should recognize that major projects are different enough from individual investigator awards—in nature (larger scale and longer term), degree of impact on a research area, and visibility at the local level—that review and selection procedures must be followed more consistently and decisions must be justified publicly. These differences are becoming even greater. Major projects are growing larger (often involving one-of-a-kind national undertakings); each directly affects the NSF budget for its immediate field of science. Some are large enough to affect the allocation of NSF's research budget across fields if annual budget increases are smaller than anticipated.

Not only are large projects costly in current dollars; they also reduce future flexibility within their fields to respond to new research opportunities. Most ideas for major projects and facilities arise from the accumulated findings and insights of individual investigator-initiated research projects. It is important for the future health and vitality of a field that the resources for large multiyear projects are not so great as to reduce the "seed corn" for future major projects. Large projects aimed at realizing current new opportunities always introduce a degree of rigidity in planning and programming that reduces overall flexibility to take advantage of future developments. It is important, therefore, that decisions to engage in large projects and their planning and implementation be closely scrutinized and fully justified.

Although controversial decisions have been relatively rare, they have revealed problems in NSB and NSF policies and procedures that could be avoided. When such problems occur or are believed to occur, they undermine the confidence in the merit review system of the research community, research institutions that compete or hope to compete for major awards in a fair process, and Congress. So far, the success of the merit review system has helped insulate NSF and NSB decisionmaking on major awards from congressional intervention. If confidence in the system is not maintained, the temptation for research institutions to try to have Congress preempt NSF decisionmaking will increase, and to the extent that legislative involvement replaces merit review with political considerations in project selection, the quality of the nation's research system may be negatively affected.

The panel recommends a number of changes to strengthen or improve the planning, review and selection, and subsequent renewal of major awards. Detailed recommendations are discussed in subsequent chapters, but the key points follow:

Clear Rules of the Game

The "rules of the game" (i.e., the criteria, procedures, and roles of participants in the merit review process) must be absolutely clear in advance.

In some cases, the criteria or requirements needed to meet them have not been clear or were seemingly redefined during the review process. Although too much detail in specifying criteria might limit the flexibility to respond to innovative proposals, we concluded that to increase procedural fairness, NSB and NSF should be more precise about the criteria and review process to be used. In particular, the primary technical criteria as distinct from the other criteria to be considered in the merit review process should be identified in advance in each case.

The panel recommends stronger planning efforts that would help contribute to clearer criteria (Recommendation 1). The panel also recommends that NSF concentrate more effort in designing a better— understood review process for each major award (Recommendation 8).

Primacy of Technical Merit

Technical merit must be the primary consideration in making awards.

The panel strongly supports the primacy of technical merit in the selection of major projects (Recommendation 3), and it endorses the use of a two-phase review process that would clearly indicate the ranking of projects on technical merit before other merit factors are considered (see next section).

Technical merit must be paramount to maximize the likelihood that the project will achieve its substantive research goals. Other criteria of merit should also be given due consideration in selecting the overall winner or winners, but any project receiving an award should rank among the very highest in technical quality. That should be made clear to all reviewers and decisionmakers, along with a sense of the nature and relative priority of each of the criteria.

NSF and NSB must be clearer in each case about the relative priority of the various criteria used, especially of the technical relative to the nontechnical criteria. Otherwise, the weightings of criteria are implicit and can shift continually at the discretion of individual reviewers and program staff.

Appropriate Roles of Peer Reviewers and Staff

The review process must be structured so that the roles of peer reviewers and staff in evaluating and recommending proposals are clearly understood, and trade-offs among technical and other criteria are clearly explained, at each subsequent level of decisionmaking.

Currently, the summary rating and ranking of proposals by staff at various decision points does not always distinguish peer review from staff judgments. Although staff should make their best case for a recommended decision, the NSF director and the NSB should always know the results of the peer review.

The two-phase review process, properly documented, would make it easier to implement this objective (Recommendation 6). This two-phase process would facilitate the preparation of a summary document that explains the rationale for the decision, including the treatment of peer review results and the trade-offs made between technical and nontechnical criteria in reaching the final decision (see next section).

Public Documentation of Decisionmaking

There should be a public document explaining the results of the review and the rationale for the final decision by the NSB.

NSB minutes rarely record the basis for a major award decision, and no public document of explanation for the final decision is prepared or disseminated. The lack of such documentation leads to public confusion and controversy that could be avoided.

The panel recommends a short, carefully prepared memorandum that summarizes the results of each stage of the merit review process and outlines the rationale for choosing a winning proposal (Recommendation 9). Such memoranda would increase public understanding of major award decisions and therefore enhance public confidence in the system that produces them.

More Stringent Setting of Priorities

Decisions to solicit proposals for very large major awards should take into account their impact on NSF's overall program as well as on the particular research field involved, and they should be contingent on the realization of expected funds and technological progress.

Careful front-end planning, combined with broad consultation with affected research communities and constant evaluation of priorities at each decision point, must be a part of the process of soliciting and reviewing proposals for a very large major award. Solicitations for awards that have serious long-range budget implications must be based on a broader range of considerations than in the past. The priority within a given field should be clearly established and compared with the overall priorities of NSF across fields. After initial approval of a large project, contingency plans for possible unfavorable program or budget developments should be made for each project and updated annually. The potential impact on NSF priorities if there are unrealized budgetary expectations or unexpected technological problems or opportunities should be carefully reviewed at each decision point. In this way, NSF and NSB would avoid letting a series of small decisions in the development of a major project result in a project that no longer matches the agency's overall program priorities or budget.

The panel calls for stronger planning efforts, including contingency plans for lower funding levels than expected (Recommendation 1), based in part on a broader range of input from research communities affected directly and indirectly by a major project (Recommendation 2). NSB should also put more emphasis on its long-range planning and priority-setting activities (Recommendation 7), and should periodically reconsider the contribution of every project to agency priorities as part of a systematic renewal process (Recommendation 10).

In making the recommendations to carry out these goals—better planning, more explicit processes and criteria, more standardized procedures, and more public justification of award decisions when made—the panel does not mean to imply that NSF and NSB are not making any attempts to meet them. Elaborate planning procedures exist (described in Chapter 2), but we believe they could be improved in certain ways to ensure that the research community is well informed and reasonably supportive, and that greater attention is paid to worst-case budgetary and technical scenarios. The project development plan requirement that exists for large-scale facility construction should be extended to all major award projects.

NSF has recently augmented its procedures for planning the review process prior to soliciting proposals (Chapter 4); the next step is to ensure that the key features of the review plan are published explicitly in the proposal solicitation document. A distinctive multistep review process has evolved for some types of major awards (described in Chapter 3); the benefits of this decisionmaking approach could be usefully applied in all major awards. In recent years, NSF has begun to include requirements for recompetition with explicit timetables in some major awards (Chapter 5); this approach could also be used more generally.

The greatest change in current policies and practices would come in the area of explicit justification of award decisions after final approval. In following the traditions of peer review of individual investigator research projects, the results of peer reviews and rationale for the award decision have not been made public (Chapter 4). We believe that such public explanation is justified by the scientific importance, budgetary size, and public visibility of major awards and would increase support for the decisionmaking process.

The recommendations in this report are offered to strengthen and protect the use of merit review in government research decisionmaking because, despite its shortcomings, it is the best way known to ensure quality. We do not think it is possible, however, to devise a perfect review process. There will always be losing applicants for a key award who would not be happy even if a new system were devised weekly in response to complaints. If our recommendations are adopted, we believe that the major award review system will still have adequate flexibility to respond to the variation and unpredictability of the real world.

The panel understands that its recommendations cannot guarantee a perfect result or prevent individuals and institutions who are denied awards from complaining about the system. This is especially true of awards for large, one-of-a-kind national facilities that must satisfy many expectations. We believe that the changes recommended in this report will result in a fairer and more understandable process; will increase confidence in, and support of, the system by fair-minded participants and interested groups; and will help forestall outside pressure to fund projects on a nonmerit basis.

2 Planning Major Projects

Awards for major projects pose some issues that are different from, and more complicated than, those involved in traditional research project grants to individuals and small groups. Major awards are much more expensive and longer term than individual research grants. Because of their size and duration, they promise to have significant impacts on the way science is funded and conducted in their field of research. Since the budget of the National Science Foundation (NSF) cannot meet all the needs of the many fields of science and engineering, each major award entails a critical initial investment decision and a subsequent long-term "mortgage" on NSF funds. If these issues are addressed fully in the planning phases, even before proposals are solicited, many problems in the review and award phase of a major project can be avoided or corrected.

BACKGROUND: PROJECT PLANNING AND BUDGETING AT NSF

In assessing NSF's planning of large projects requiring major awards, the panel identified certain essential features of an ideal planning process. Planning for major research projects should be an integral part of overall planning and setting of priorities at NSF. Priorities should be based primarily on research opportunities. Major projects should be weighed against individual research projects and other modes of research support in deciding on the best overall program. Plans for major projects should be extensively reviewed and widely discussed with advisory committees and the research community before funding is approved. Funding decisions should be based on the best estimate of operating and maintenance costs over the expected lifetime of the project, as well as initial setup or construction costs. The conditions under which the funding award would be renewed, recompeted, or terminated should be considered in the planning process and clarified up front, in case the project did not go as planned or the anticipated budget resources were not realized.

Long-Range Planning at NSF

NSF has a number of arrangements for obtaining advice from, and consulting with, the science and engineering research communities about its programs and projects both large and small. Taken together, they form a continuous, decentralized, and open planning process that may be "driven by a scientific breakthrough, the availability of a new technology, national or international concerns, or simply the existence of a new idea" (NSF, 1990a:3). NSF also participates in the annual federal budget process, which involves top-level decisionmaking across research fields and agencies. Although the budget process is an annual exercise, the resulting decisions may have much longer-term implications.

1. Advisory Committees: NSF has had 96 formally chartered advisory committees with nearly 6,500 members from the science, engineering, and education communities. Those committees include standing panels that programs in some directorates use to review proposals and, until recently, programmatic advisory committees to many of the program divisions.

Most divisional advisory committees represented a particular discipline or scientific field, although some were more broadly based to advise on interdisciplinary programs. They met once or twice a year in open meetings. NSF consulted the advisory committees in the development of its program plans and priorities, and involved them

PLANNING MAJOR PROJECTS 41

in assessments of current activities. Occasionally, they reviewed proposals for major awards (the Materials Research Advisory Committee, for example, reviewed the National High Magnetic Field Laboratory [NHMFL] proposals).

In response to the recent executive order directing each federal agency to cut advisory committees by a third, NSF is eliminating 34 of its advisory committees. That is being accomplished by eliminating division-level committees and establishing advisory committees to each of the six research directorates[1]. This drastic reduction in programmatic advisory committees will allow NSF to retain most of the proposal review panels, although some of them may be consolidated at the directorate level. It will also affect the nature of advisory input on program issues from the scientific and engineering communities, and will reduce interactions with the disciplines, although it may promote interdisciplinary planning.

2. <u>Quarterly Reviews</u>: The NSF director and senior staff meet with each directorate four times a year to review overall program activities, budgets, and management. Division (now directorate) advisory committee members and other outside experts are usually invited to participate. The round of reviews held each spring provides part of the input into the long-term planning session of the National Science Board (NSB) in June. According to NSF, these sessions "often provide the first airing of concepts which later emerge as scientific initiatives or new programs," including those destined to be major awards (NSF, 1990a:5).

3. <u>Ad Hoc Task Groups</u>: NSF may form an ad hoc advisory group or interdirectorate staff working group to assess new program ideas, whether they involve large or small projects, and to consider how they might be implemented. NSF's first large projects—the radio telescopes at Greenbank, West Virginia, and optical telescopes

---

[1] A seventh committee will advise the Education and Human Resources Directorate.

at Kitt Peak, Arizona—grew out of discussions at scientific conferences funded in the early 1950s by NSF (England, 1982:280-281). In more recent examples, an NSF-appointed ad hoc advisory panel proposed and developed the specifications for a next-generation national high magnetic field laboratory, which was awarded in 1990 (NSF, 1988a). Another NSF-appointed panel on large-scale computing in science and engineering called for increased supercomputer access by academic researchers (NSF, 1982), and a staff working group developed an action plan that included what became the supercomputer centers program in 1984 (NSF, 1983).

4. Professional Societies: Representatives of scientific and engineering societies and associations participate regularly in advisory committee and NSB meetings. The NSF director and other senior staff meet with officials of such groups and with other groups of higher education officials, industrial and federal laboratory directors, and foundation heads. NSF staff also stay informed about scientific developments through participation in activities of professional societies.

5. National Research Council: NSF often asks the National Research Council or one of its governing academies (National Academy of Sciences, National Academy of Engineering, Institute of Medicine) for scientific, policy, and programmatic advice. These reports can have a major influence on program development and decisions on priorities. National Academy of Engineering reports helped shape the Engineering Research Centers program, for example. Other examples include the "decade studies" of astronomy and physics that have been influential in setting priorities among major NSF-supported projects in those fields, particularly in establishing the categorization of activities among which priorities are set.

In addition to the external advisory mechanisms listed above, the NSB, which has statutory responsibility for setting NSF policies and approving its programs (and for reviewing and approving the major individual awards addressed in this report), has members who represent many of the scientific and engineering fields supported by NSF. The NSB assesses current and planned NSF activities through its standing committees on Programs and Plans and on Education and Human Resources—and through special task forces, committees, and commissions. Finally, one-third of the professional program staff of NSF consists of visiting scientists and engineers on one- to three-year assignments, who bring another source of knowledge about current research opportunities and needs in their field. Rotation also helps bring in new ideas and perspectives.

Annual Budget Process

The ongoing decentralized activities described above result in many interesting suggestions, some of them for major large-scale projects. In some cases, these activities yield broad and clear consensus on priorities among major projects in a field. For example, the decade studies have helped forge consensus on needs and priorities in astronomy. Another example is the traditionally "small-science" earth sciences community. After a series of planning discussions over several years, that community agreed that a large-scale coordinated program had become necessary to make progress. The result was the Continental Lithosphere program, which included small-grant research, coordinated field projects, continental drilling projects, and a global network of seismographs.

Not all ideas that turn into major awards originate from the bottom up. The Engineering Research Centers initiative, for example, came out of discussions between NSF officials and the Science Advisor to the President. It was seen as a response to the declining economic competitiveness of the United States as well as a way to meet an engineering research and education need; that initiative helped justify a doubling of the NSF budget at the same time that it tried to help reform academic engineering education by introducing more experience in interdisciplinary work and team efforts.

Although NSF's internal planning and programming are continuous and long term, it is funded only one year at a time. Decisions about which existing activities will be continued, expanded, scaled back, or terminated and which new initiatives will be funded for the first time in the coming fiscal year have to be made with some uncertainty about the current year's budget and greater uncertainty about future budget levels (the FY 1993 budget process is described in Box 2-1). Out-year adjustments in the budget category for traditional individual investigator grants are relatively easy to make because the grants average 2.5 years in length; thus, about 40 percent of that part of the NSF budget becomes available for reprogramming each year. Major projects require longer-term funding commitments that cannot be reduced as easily if the NSF budget is cut or fails to grow as much as anticipated in future years. As a result, increased funding for individual investigator grants tends to be squeezed out first unless there is careful contingency planning.

MAJOR PROJECT PLANNING AND BUDGETING

In addition to the long-range planning exercises and annual budget process described above, the NSB has special review and approval procedures for most if not all activities expected to result in major awards. According to policies adopted by the NSB in January 1979, the NSF directorates are supposed to submit for approval by the director and the NSB "project development plans" for "big science" initiatives (NSB, 1979). Big science projects are defined by NSB as those that have certain characteristics:

- large-scale commitment of financial resources;
- investment of capital in facilities and major equipment;
- duration of several years or more; and
- continuing expenditures for maintenance, replacement, operating costs, and research.

BOX 2-1: A Recent Budget Cycle

The budget process for FY 1993, which was typical of the annual budget process of the National Science Foundation (NSF), proceeded as follows:

Intra-NSF Planning Period. The budget process for a fiscal year begins nearly two years before the start of that fiscal year. For example, NSF began planning its budget for FY 1993 in early 1991 at a retreat attended by the director, deputy director, assistant directors, and other senior NSF staff. The group reviewed the current five-year plan and developed a list of possible issues and initiatives for the upcoming budget cycle.

During the same period, each of NSF's research directorates underwent planning exercises involving its divisional advisory committees. The results were presented to the director as part of the spring round of quarterly reviews. These activities included a workshop held in early 1991 on cognitive science to lay the groundwork for a formal NSF-wide initiative; a proposed initiative on integrated manufacturing systems of interest to the White House; plans to implement State University/Industry Cooperative Research Centers; and a nonlinear science initiative. The directorates with large investments in facilities were concerned about capital budgeting.

In May 1991 the director, assistant directors, and the National Sciences Board (NSB) met with NSF advisory committee chairs to go over the issues and potential initiatives emerging from the planning process. A few themes began to emerge. Activities of the Federal Coordinating Council for Science, Engineering, and Technology (FCCSET) were becoming more important, and all directorates wanted to be involved or to increase their participation if already involved. It was agreed that more centers were desirable, although no major initiatives were proposed. There was a strong desire to strengthen disciplinary research activities by the size, duration, and budget share of awards. It was decided not to handle capital budgets as an NSF-wide issue, but to leave them within the directorates so that they could be closely linked to the needs of each field.

June Long-Range Planning Meeting. The next major internal planning event was the June 1991 meeting of the NSB, which is traditionally devoted to long-range planning. The director presented staff recommendations in the context of a newly prepared five-year plan for FY 1993-1997. The NSB in turn identified activities it wanted to expand

Box 2-1 continued

or develop, and provided general guidance to the staff on the size and scope of the budget proposal for FY 1993.

The NSB reached general consensus that the FY 1993 budget request should emphasize core disciplinary research and investment in education, increase the average length of awards from two to three years, continue strong participation in the three FCCSET initiatives, expand programs in biotechnology and advanced materials that were expected to become new FCCSET initiatives in FY 1993, begin new initiatives in advanced manufacturing and environmental research that were expected to become FCCSET initiatives in the future, and hold a competition for several new Engineering Research Centers (ERCs) in strategic research areas.

The budgetary context for these deliberations was as follows: the FY 1992 budget was under consideration by Congress during the spring of 1991. Despite an administration commitment to double the NSF budget between FY 1987 and FY 1992, its budget request for FY 1992 amounted to only a 67.7 percent increase over FY 1987. Budgetary growth had been concentrated in Education and Human Resources (EHR) programs. The FY 1992 request for these programs was 256 percent more than its appropriation in FY 1987; the Research and Related Activities (R&RA) budget request was just 48.5 percent higher than that of FY 1987. In turned out that the administration's requested budget increase of 17.5 percent for FY 1992 was not approved in full by Congress. The actual overall increase eventually approved by Congress (in nominal, not constant, dollars) was 11.1 percent (10.6 percent for R&RA, 44.4 percent for EHR).

NSF had begun a number of new major research and related projects during this period, including 10 ERCs and 25 Science and Technology Centers. It was also continuing large projects begun before 1987, such as the Very Long Base Array radio telescope facility; an upgrading of the radio telescope at Arecibo, Puerto Rico; four Supercomputer Centers; and ERCs. The FY 1992 budget then being considered by Congress contained a request for $51.5 million to continue developmental funding of the two 8-meter optical/infrared telescopes; continued construction of the National High Magnetic Field Laboratory (NHMFL); and initial funding to begin another major construction project—the Laser Interferometer Gravitational Wave Observatory (LIGO).

Box 2-1 continued

Preparing the FY 1993 Budget Estimates. Between the June 1991 NSB planning meeting and September 1, 1991, when the NSF budget request was submitted to the Office of Management and Budget (OMB), the preliminary allocation of resources to individual directorates and divisions took place. The NSB reviewed and approved the budget request at its August meeting.

During September, OMB and NSF assistant directions held a series of meetings at which NSF clarified its requests and responded to questions. NSF-OMB interaction continued throughout the fall. In late November, NSF received its "passback" budget figure from OMB and began formulating the budget request to Congress.

NSF was also involved in the interagency planning process under FCCSET. NSF identified funds in its budget request for FCCSET initiatives totaling $948.6 million (up from $415.5 million in FY 1992). These initiatives were the Advanced Materials and Processing Program; Biotechnology; High Performance Computing and Communications; Mathematics and Science Education; and the U.S. Global Change Research Program.

As a budgetary strategy, the discussion of FCCSET initiatives was more prominent in the final budget request than it had been in the budget developed during the summer. Decisions were made to continue to build up materials, biotechnology, and environmental research as part of new or future FCCSET initiatives. At the same time, NSF strongly emphasized the similarity between the types of research supported by FCCSET initiatives and the traditional disciplinary research programs. It also stressed the importance of individual investigators in all types of NSF support. Centers were no longer emphasized, and new center initiatives were modest: $4 million for two new ERCS and $3 million for a national center for ecological synthesis and analysis. Some new disciplinary initiatives were included although they were not discussed at the NSB or NSF-wide level (e.g., a new hydrological sciences program); others (e.g., the nonlinear science initiative) were dropped.

The president's request for R&RA for FY 1993 was $2.21 billion, compared with $1.88 billion recently approved by Congress for FY 1992. It included $79.0 million for continued construction of major facilities: LIGO ($48.0 million); the two 8-meter GEMINI telescopes ($17.0 million); and the NHMFL ($14.0 million). The NSF budget request also would have increased funding for research centers by 9 percent to $153 million, including two new ERCs.

> Box 2-1 continued
>
> *Congressional Review.* Although NSF's budget grew between FY 1987 and FY 1992, it did not double as planned because NSF was competing with other programs, including housing and veterans affairs. Also, Congress set different priorities among NSF programs. The R&RA part of the budget, which had been growing at a slower rate than other NSF accounts, barely increased between FY 1992 and FY 1993.
>
> NSF is not always able to distribute reductions in the requested budget in accord with its own program. For example, Congress approved $38 million for LIGO, but NSF was directed to spend the full $43 million initially requested, with the $5 million difference to come out of funding for other physics projects (this mandate was later lifted when the extent of its impact on other NSF programs became clearer).
>
> Within NSF, the budgetary emphasis on human resource programs and programs related to economic competitiveness has affected other parts of the NSF portfolio over time (NSF, 1990a:8). One result has been the persistent underinvestment in secondary priorities, such as the acquisition and maintenance of research equipment and facilities. This, in turn, has resulted in deteriorated facilities and deferral of planned facilities and centers.

According to the NSB, project development plans for big science projects are supposed to document the following:

- scientific need;
- views of the appropriate advisory group concerning the priority of the project; its effect on the balance and concentration of "big science versus little science" within the field, under varying resource assumptions (including essentially level budgets); and the opportunities that would be forgone by undertaking or not undertaking it;
- estimates of all initial and out-year costs;
- principal management, procurement, and legal considerations;
- origin and periodicity of management and fiscal reports, and timing and other considerations for evaluating the project;
- identification of principal phases or milestones; and
- arrangements to update plans at least annually.

All such big science projects proposed for a given budget year are to be reviewed at the same time so that NSB can set priorities (based on prior review by, and recommendations of, the director). Once approved, project development plans are supposed to be updated annually and revised whenever a significant shift in the terms or funding of the project is being considered. The approved plan is also the basis for dealing with the Office of Management and Budget (OMB), Office of Science and Technology Policy (OSTP), Congress, and other external organizations and groups.

According to the NSB policy document, the size of a project must be considered both in absolute (cost and complexity) and in relative (the share of resources of a particular field and of NSF's overall budget) terms. The policy does not set a limit on the share of a research program's budget that can go to large-scale projects, because the appropriate balance varies across fields and is subject to periodic review by advisory groups, staff, and the NSB. If funding is too high to be accommodated within the appropriate disciplinary budget, the director and NSB must decide whether or not the project can be accommodated within anticipated NSF budget levels or must become a special item justified above and beyond anticipated NSF budgets.

Capital Facilities Planning

In 1983, the NSB planning and budget committees examined NSF's capital facilities planning and budgeting procedures (NSB, 1983c:9). That review concluded that NSB's 1979 policies on initiating large-scale projects and its regular long-range planning procedures were adequate. That stance began to change in 1988, when a 10-year projection indicated a need for $1.6 billion in capital to repair and replace aging NSF-supported research facilities and enable them to take advantage of new research opportunities (NSB, 1988b). That amount was well beyond the level that could be met even by the planned doubling of the NSF budget. The projection indicated a need to treat major capital items of $10 million or more differently in order to improve planning, decide on priorities, decide on funding alternatives, and justify them to OMB and Congress.

By FY 1990, NSF's support for major research facilities, (e.g., astronomical observatories, supercomputing centers, high-energy and nuclear physics facilities, oceanographic research ships, and atmospheric research facilities) totaled almost $400 million, 25 percent of its appropriation for Research and Related Activities. About one-quarter of that was for capital costs. Three-quarters was for operations, including research. NSF estimated that support for facilities would increase to about $620 million by FY 1995 (NSF, 1990b:1).

As a result of these trends, NSF adopted a new approach to budgeting for capital facilities in the Mathematical and Physical Sciences (MPS) Directorate (NSF, 1990b:6):

> Historically, both capital and operating costs for a facility have been included within the appropriate discipline's budget. However, the incremental funding required for major capital items is often too large for a single discipline's budget.
>
> A new subactivity, Major Research Equipment, was established in FY 1990 to support the construction of new, high-priority large-scale research facilities in the physical sciences. Projects budgeted with the Major Research Equipment subactivity are LIGO, NHMFL, the Green Bank Telescope, and the two 8-meter telescopes. All other specialized research facilities continue to be budgeted within the disciplines. The availability of sufficient funding for operations for both new and existing facilities is of increasing concern, and is being specifically addressed in NSF's long-range plans.

The success of the capital facilities subaccount in MPS remains to be seen. It has been used to justify and gain support for very large facilities each involving more than $100 million over its lifetime (including non-NSF funding). NSF was able to start the major projects it had proposed, including the 8-meter telescopes, NHMFL, and LIGO. NSF also may succeed in establishing the account as a revolving fund with Congress, enabling the agency to fund new projects as approved ones are completed and turned over to the research programs for operating support.

One risk of putting the capital projects in a separate account is their increased visibility to Congress. Congressional committees may direct that projects in the subaccount be fully funded even when NSF does not receive its full budget request. So far, that has not happened. In FY 1992, when it did not receive its full request for Research and Related Activities (R&RA), NSF was able to revise its operating plan by reducing the funding for ongoing construction projects from $51.5 million to $37.9 million. In FY 1993, however, when NSF received about the same budget for R&RA as in 1992, it was directed at first to protect funding for LIGO even though it wanted to scale back this activity. NSF had to negotiate with the appropriations committees to scale back LIGO funding in FY 1993 so that it did not squeeze funding for other research. Thus, whereas NSF may gain the opportunity to justify expensive new facilities projects, it also may lose the flexibility to adjust expenditures within a research field in the face of reduced funding, especially when Congress does not follow NSF's capital budget plans.

FINDINGS AND RECOMMENDATIONS ON PLANNING AND BUDGETING

Findings

The NSF planning process is very decentralized, continuous, and open. It works well to encourage new ideas. The process works best within individual fields or disciplines (e.g., physics, astronomy, biology)—but not across them. The budget process also imposes a very short-term planning horizon, because NSF receives annual, not multiyear, funding.

Because the planning process produces more good ideas than can be funded in any given year, the NSF director—with the advice of the assistant directors—makes the key internal decisions on priorities, subject to NSB oversight and concurrence. Initiatives and priorities do not always come from the bottom up; higher-level governmental oversight groups also affect priority decisionmaking (e.g., the Science Advisor to the President, OMB, the Federal Coordinating Council for Science, Engineering, and Technology, and congressional committees).

Although a large share of NSF's resources goes into building, maintaining, and updating large-scale facilities and instruments, capital budgeting is currently done within fields if at all. Most capital projects are in two directorates, MPS and Geosciences. MPS recently established a budget subaccount for "Major Research Equipment" that includes LIGO, the GEMINI telescopes, and the National High Magnetic Field Laboratory.

Recommendations

Project development plans are usually prepared for projects involving the construction of facilities (e.g., Continental Lithosphere Program (which included the Incorporated Research Institutions for Seismology [IRIS]), 1984; LIGO, 1984; NSFNET, 1987; NHMFL, 1988; 8-meter GEMINI telescopes, 1991). Project development plans have not been prepared for centers or center programs (e.g., Earthquake Engineering Research Center, the Engineering Research Centers [ERCs], or the Science and Technology Centers [STCs]). In those cases, NSB approved the next stage, the proposed solicitation document. Project development plans also are not required for ongoing campus-based research facilities primarily intended for local use (e.g., nuclear physics facilities at the University of Illinois, Michigan State, and Indiana University).

The decision to construct a major research facility, launch a program of interdisciplinary research centers, or support a large-scale coordinated research program has a strong and lasting impact on the way research is carried out in a scientific or engineering field. Also, the large cost of major projects imposes opportunity costs—the funds committed to them, usually for long periods, cannot be used for other, perhaps more productive, but less visible, activities. These opportunity costs are felt strongly within the affected research area, and in this era of very constrained federal budgets, they may affect other areas as well.

PLANNING MAJOR PROJECTS

The case studies provide illustrations of these effects. The decision to establish a national Earthquake Engineering Research Center had a major impact on the mode of earthquake engineering research supported by NSF, because it shifted one-third of the approximately $15 million a year in NSF's earthquake hazard mitigation program from individual investigator grants to the center mechanism. LIGO, the new $200 million instrument for the specific research area of gravitational physics, is a relatively costly project that may have an impact on the funding of other fields as well as physics if NSF's research budget does not increase as much as planned.

The NSB recognized the critical importance of the initial decision to undertake a major project and since 1979 has required submission of a formal project development plan for the construction of large research facilities. This requirement has not been followed in all major award cases, however, although the need for it today is greater than ever before, for several reasons:

- Scientific opportunities are growing faster than NSF resources, which means that costly new initiatives need to be subjected to special scrutiny to ensure cost-effectiveness in comparison with alternatives such as a program of small grants.

- The scale of research has increased in size and cost, bringing big science into additional research fields, which means that the interrelationships among modes of research within a field and with closely related fields should be carefully worked out to ensure balance.

- The more needed and better planned the project is, the easier it will be to solicit good proposals, develop appropriate review criteria and procedures, and identify the best proposal to carry out the goals of the project.

The panel concluded that diligent use of the project development plan mechanism for all major awards, not just for facility construction projects involving construction of facilities, would help ensure that such projects are well justified scientifically, of high priority, well designed and technically feasible, and complementary to the rest of NSF's activities in that and related fields. Wider use of project development plans would also ensure that appropriate analyses of options under different budgetary and technological scenarios have been conducted.

Recommendation 1: Justification for Major Project Awards

The NSB should ensure that the large-scale research-related projects that result in major awards are well justified and planned—that is, each is (a) scientifically justified, (b) technically feasible, (c) designed to enhance other activities already in place to achieve the proposed project's goals, (d) of high national priority, and (e) the subject of careful contingency planning.

These factors should be fully considered before proposals are invited and issues of appropriate review procedures and criteria are addressed:

a. *Scientific Justification*: A decision to undertake a major project must promise important scientific contributions, whether directly through support of large-scale or interdisciplinary research that could not be done otherwise or indirectly by providing access to state-of-the-art facilities for individual researchers or research groups.

b. *Technical Feasibility*: Major projects must be technically "doable" as well as scientifically desirable. In the case of state-of-the-art facilities, research and development on new instruments should be advanced enough to justify full-scale deployment. In the case of ERCs and STCs, there should be evidence that they will be able to produce research that will be useful economically.

c. *Complementarity*: Major projects should be part of, and should enhance, a balanced program of mechanisms for supporting productive science and engineering research and education in a field, which work together to maximize scientific progress in that field.

There is no a priori way to determine what share of NSF's resources should go to individual investigator and small-group research projects; university-based, regional, or national facilities; centers; or large research groups—the balance varies by field and within fields over time. Major projects reduce NSF's overall programmatic and budgetary flexibility. The burden of proof, therefore, should be on a potential major award to demonstrate that it might add more value than an equally expensive program of individual projects or other, more flexible research mechanisms. Also, NSF should take steps to ensure that alternative views are heard in the final decisionmaking. That might involve building in an advocacy process and would help ensure that the best case is made, for example, for an equivalent program of individual investigator or small-group grants, or for instrumentation grants to many or several colleges and universities rather than for the establishment of a single national facility.

d. *Programmatic Priority*: In NSF's open and continuous planning process, more good ideas come up than can be funded. NSF leadership and the NSB must make decisions about allocating the approximately 40 percent of NSF's budget base that is available each year (the rest is committed to multiyear projects), as well as any overall budget increases or decreases. These decisions are difficult because they inevitably favor one research area over another. The more distant the areas are from each other, the less relevant are purely technical considerations in making the appropriate choices.

According to NSB policy, all large projects subject to the requirement for a project development plan are supposed to be reviewed at the same time so NSB may establish priorities among them. This procedure takes place annually at the June NSB meeting, which is devoted to long-range planning, but the full set of project development plans is not always available to provide detailed information on all factors that must be taken into account in setting priorities (as intended by the project development plan policy).

e. *Contingency Planning*: The broader implications and long-term budgetary impact of major awards should be carefully considered, including a worst-case analysis, and weighed against the project's need and priority, as well as its risk and potential payoffs.

Large projects tend to achieve tremendous momentum, scientifically and politically. If the NSF research budget fails to grow or shrinks in real terms, large projects may squeeze out other less visible, but perhaps equally valuable, programs. This analysis should be ongoing, with specific trigger points identified when a project should be postponed, scaled back, stopped, or stretched out in time.

NSF and NSB currently address all these planning criteria in one way or another and to some degree. The panel would like to see them addressed more systematically by adoption of the format for project development plans for all major awards (not just those involving construction). In some ways, the criteria in Recommendation 1 go beyond those currently required in a project development plan, especially in seeking a broader context for decisions and contingency plans. A major project not only should be justified as part of an overall plan within its field of research, but also should be considered by NSB along with all other major awards vying for funding during that budget cycle and in longer-range plans. The consequences of smaller-than-expected appropriations should be very seriously considered, and contingency plans made and communicated to the relevant public, including congressional committees, as early as possible. The contingency plans should include explicit trigger points for reconsideration of projects that develop technical or budget problems (or opportunities that justify additional investments).

The additional emphasis on national priorities and contingency planning requires consultation with a wider range of research interests than before. Increased competition for limited resources means that unanimity in a particular field (even if it can be achieved) is not a sufficient condition for going ahead with a project. A broader perspective on research priorities is required than before. The NSB itself should place greater emphasis on its responsibilities for planning and setting priorities (see Recommendation 7) and should look for a wider base of consultation early in the development of major projects.

Technical feasibility is another requirement highlighted here. In the 1979 NSB policy on project development plans, it appeared as a "probability of success" subfactor under the "scientific need" requirement. Facilities and centers pose the issue of technical feasibility more sharply than do individual investigator grants. Before a telescope, accelerator, or other facility can help researchers, it must be built to very exacting specifications. Meeting the specifications to do forefront research involves expertise and judgment about engineering feasibility and accuracy of cost estimates.

This expertise and judgment is different from that necessary to determine the scientific promise of individual investigator grants. For example, an issue in the LIGO project, in addition to its affordability at a time of slowly-growing budgets, was its technical readiness. Achieving the goals of the project involves a degree of instrument sensitivity several orders of magnitude beyond that previously reached. Cost estimates had already doubled several times during the planning of the project. Technical uncertainties and increasing cost made it an especially risky decision.

Center programs pose a different technical feasibility issue. In most cases they are created to produce research that promises to be economically beneficial. Determining whether their research areas are economically relevant involves different expertise than that required to evaluate whether the research will also be of high quality.

These initial planning criteria are focused on a critical decision—whether or not to initiate a major project in the first place. They are also important considerations in reviewing the eventual proposals, and the review process and reviewers should be selected accordingly (although the emphasis in selecting the winning proposal shifts to technical merit). The criteria and process for selecting proposals are addressed in the next chapter.

Recommendation 2:   Involvement and Support of the Research Community in Planning

The NSB and NSF should make stronger efforts to see that the basis for initiating large-scale activities is well explained, understood, and accepted to the extent possible by affected research communities. NSB and NSF should take steps to ensure broader consultation with relevant communities beyond those benefiting directly from a major project award, including educational, governmental, and industrial organizations and institutions.

Greater understanding of, and support for, the goals of a major project by the research community makes it more likely that high-quality proposals will be submitted, that external peer reviewers and advisory groups will understand and use the criteria appropriately, and that the final award decision will be understood and accepted, although NSB and NSF should not always wait for unanimity in the research community before proceeding. Even in the latter case, NSF should make every effort to inform the relevant affected communities about its plans.

The panel found that the "need" for a major project has not always been understood or accepted by the relevant research community. Among the case studies, for example, more groundwork was laid for the Engineering Research Centers program than for the Science and Technology Centers program. Both were initiated at higher levels, but NSF went through several planning exercises with the engineering community before deciding to approve the ERC program and solicit proposals. These exercises included the NSB (NSB, 1983a, 1983b) and early consultation with the engineering research community (NAE, 1983). Although a project development plan was not submitted per se, the draft solicitation was based on the recommendations of a committee of the National Academy of Engineering (1984). It was submitted to the NSB with a three-page concept paper from the staff.

The STC program was launched more hurriedly in January 1987 in response to a presidential speech. The only outside input was a National Academy of Sciences committee report in June 1987 on how to design the program and the process for soliciting proposals (NAS, 1987). The NSB reviewed and approved the draft solicitation announcement at its August 1987 meeting. Although the response was huge (323 proposals), there were more confusion about and lack of support for the concept than there had been for the ERC program. In this case the problem of gaining consensus in the research community was made more difficult by the breadth of areas covered by the program; virtually every field supported by NSF was eligible to apply.

Another case—the global seismology network of IRIS—illustrates the benefits of early involvement of the research community in developing a major project. In this case, NSF worked with the relevant research community for several years in developing a balanced program for the solid-earth sciences that encompassed (1) small-project support for individual investigators and small groups; (2) large multi-investigator, interdisciplinary projects; and (3) major facilities and instrumentation for crustal drilling and seismology. The program was presented to NSB in a project development plan that included all aspects of the program—small-grant support as well as large facilities that come to NSB later for approval of individual awards. The program was implemented smoothly, and awards for several large-scale facilities were made without controversy, although the earth sciences hitherto had been primarily a small-science enterprise.

The research community is not homogeneous; it consist of many specialized, mostly discipline-based groups that have different needs and priorities. Depending on the project, it may be difficult to consult with, and gain the support of, every affected research community. Attempts to broaden the range of groups consulted also makes consensus building more difficult. The panel nevertheless concluded that it is highly desirable to involve and seek the support of the research community as much and as early as possible in the process of deciding to support a major project. Such early and continuous involvement can help ensure that such projects are scientifically justified and truly complementary to other activities in a field. Such involvement also assists in determining the priority of the project within a field. It can help ensure that the solicitation is well designed, that the external review is carried out with a better understanding of what is required, and that the final award decision is better understood and supported.

Sometimes, as in the IRIS case, NSF may be able to seek the consensus of the research community affected in deciding on new or revised programs, especially those involving major awards for facilities, centers, or other large-scale and long-term activities. NSF can and often has played a leadership role by sponsoring such consensus-building activities from time to time in each research field.

That does not mean, however, that NSF must always wait for consensus to develop in a scientific field before developing new projects and programs. The federal budget process responds to initiatives from many sources, and budget opportunities sometimes arise on short notice. In other cases, conditions seem ripe for a new field to develop in which only a few, probably younger, researchers are engaged. In these and other cases, NSF has a leadership role to play. NSF and NSB should retain an ability to initiate and fund some activities of high risk, or innovation in science will suffer. Even when they take a leadership role, however, it is important to bring along as wide a community as possible through active consensus-building efforts.

Regardless of where the idea for a major project originates, NSF should develop and communicate a well-thought-out rationale for the initiative to the affected scientific community and should involve it as much as possible in planning the project. The project development plan procedure calls for NSF to document the scientific need for the project and the views of appropriate advisory groups on its priority; its effect on the balance of research mechanisms in the field; and the opportunity costs of undertaking it.

Major projects almost always have broader effects than individual investigator awards—on colleges and universities, industry, and local economies—and the NSB should ensure that plans are communicated to these constituencies and that their views are given an opportunity to be heard and seriously considered even if they are not followed in the end. This kind of consultation is especially necessary when NSF undertakes projects aimed at applications and national goals or those involving nonfederal cost sharing, as in the ERC and STC programs.

Generally, NSB should be very wary of approving projects in which community input has not been considered and documented. That does not mean NSF must wait for universal consensus within a scientific field before moving ahead; indeed, initiative by NSF leadership and NSB may be appropriate. However, if a new initiative is undertaken without substantial consensus, the rationale for the project and reasons for going against criticism should be explained clearly in public documentation (see Recommendation 9).

3 Awarding Major Projects: Criteria and Review Procedures

The underlying theme of this chapter is that although each major project is different at some level, common characteristics of major awards distinguish them from individual research grants. These characteristics justify special review policies and procedures that are more explicit and consistent across the National Science Foundation (NSF) than is necessary for small research grants. Major projects:

- represent a substantial investment of NSF resources in a research area, with greater political costs and financial liabilities in the case of failure;
- are more complex than traditional individual investigator or small-group research projects and involve more varied criteria, some less exclusively scientific and technical than those of individual investigator proposals; and
- promise a substantial impact not only on their research community but also on the institutions and localities in which they are located.

This chapter begins by providing background on the history of merit review (and the role of peer review within it) at NSF. It then addresses and makes recommendations concerning two parts of the proposal review and award process for major projects: the review criteria used to select the most qualified proposals, and the procedures to ensure that the review criteria are appropriately applied and documented.

BACKGROUND:
THE MERIT REVIEW PROCESS AT NSF

The policies and procedures for handling major awards today have evolved from, and closely resemble, those used for grants for small research projects. External peer review has been used from the beginning at NSF to evaluate all research and education proposals. The first set of research grants was approved by the National Science Board (NSB) and awarded in 1952 after the following process: "an independent evaluation of each of fifty-eight proposals by at least three expert reviewers; a two-day meeting of a screening panel of eleven consultants expert in the fields represented; a meeting of the newly-formed Divisional Committee regarding general policy and program within the field of the Division; and consideration by the full staff of the Director" (England, 1982:166, quoting NSB minutes of February 1, 1952).

Peer review of research proposals for technical merit by outside scientific and technical experts has been so important at NSF that, for years, the shorthand title for the award process was simply "peer review" (NSF, 1975:4). Criteria that augment technical quality and competence have become more prominent over time, however. These criteria include immediate practical relevance, and the development of science and engineering capacity in all groups and regions of the country. In 1985-1986, NSF (1986a) had an outside advisory committee study the proposal review process. As a result of the committee's report, the NSF director adopted the term "merit review" to emphasize the greater role of expanded criteria, especially for "center-based activities, research groups, and shared facilities" (NSF, 1986b:D-2).

NSF's current policy is that all formal proposals for grant funding are subject to peer review by appropriate experts external to the agency, with minor exceptions listed in a policy document approved by NSB (e.g., small travel grants, the Small Grants for Exploratory Research program). In addition, for special cases in which peer review is "impracticable or too costly," the NSF director may waive it (NSF, 1993c:I-4).[1] No such waivers have been requested in recent years (Massey, 1991, 1992).[2]

Peer reviews are advisory; that is, external experts are used by full-time NSF program officers to help them make decisions on proposal funding. However, the final decision can be based on a range of considerations. Decisions are affirmed or reversed at higher administrative levels, depending on the size of the project and other criteria. If the projects are large enough, or raise new programmatic or policy issues, they must be reviewed and approved by the NSB.

Current Review Criteria

According to the NSF *Proposal and Award Manual* (NSF, 1993c:I-3), it is agency policy to give every research proposal "proper consideration in accordance with established criteria approved by the NSB." The NSF *Grant Policy Manual*, which is published for the use of proposers, says that the review and evaluation criteria are "to be applied to all research proposals in a balanced and judicious

---

[1] External review is not required for proposals submitted in response to formal solicitations governed by the Federal Acquisition Regulations review, that is, in response to Requests for Proposals for contracts. Almost all major research awards, however, are cooperative agreements (a type of grant) rather than contracts.

[2] Some major awards have been exempted in the past. In FY 1985, for example, peer review was waived for two planning proposals for multi-institutional consortia to manage large earth sciences projects, because each of the consortia included virtually all the institutions involved and thus made it impossible to find reviewers without conflicts of interest. Instead, the projects—IRIS (Incorporated Research Institutions for Seismology) and DOSECC (Deep Observation and Sampling of the Earth's Continental Crust)—were discussed extensively by the advisory committee for earth sciences and the NSB committee on programs and plans (Bloch, 1986).

manner, in accordance with the objectives and content of each proposal" (NSF, 1989:II-6).[3]

The criteria are as follows:

1. *Competent performance of the research* relates to the capability of the investigator(s), the technical soundness of the proposed approach, the adequacy of the institutional resources available, and the proposer's recent research performance.[4]

2. *Intrinsic merit of the research* is used to assess the likelihood that it will lead to new discoveries or fundamental advances within its field of science and engineering, or will have substantial impact on progress in that field or in other scientific and engineering fields.

3. *Utility or relevance of the research* is used to assess the likelihood that it can contribute to the achievement of a goal extrinsic or in addition to that of the research field itself, and thereby serve as the basis for new or improved technology or assist in the solution of societal problems.

4. *Effect of the research on the infrastructure of science and engineering* relates to its potential to contribute to better understanding or improvement of the quality, distribution, or effectiveness of the nation's scientific and engineering research, education, and human resources base.

---

[3] NSF recently made the *Grant Policy Manual* available electronically on Internet and plans to update it periodically.

[4] The phrase, "and the proposer's recent research performance," which appears in the NSF (1993c) *Proposal and Award Manual*, is an addition to the criteria adopted by the NSB in 1981. The criteria listed on the back of the evaluation form sent to external reviewers (NSF Form 1) contains a similar addition: "Please include comments on the proposer's recent research performance."

NSF policies are not very specific about the relative priorities and weightings these different criteria should have. The *Proposal and Award Manual* (NSF, 1993c:I-3) provides the following discussion:

> Criteria 1, 2, and 3 constitute an integral set that are applied in a balanced way to all research proposals in accordance with the objectives and content of each proposal. Criterion 1, performance competence, is essential to the evaluation of the quality of every research proposal. The relative weight given criteria 2 and 3 depends on the nature of the proposed research. Criterion 2, intrinsic merit, is emphasized in the evaluation of basic research proposals, while criterion 3, utility or relevance, is emphasized in the evaluation of applied research proposals. Criterion 3 also relates to major goal oriented activities that the Foundation carries out such as those directed at improving the knowledge base underlying science and technology policy, furthering international cooperation in science and engineering, and addressing areas of national need. Criterion 4, effect on the infrastructure of science and engineering, permits the evaluation of research proposals in terms of their potential for improving the scientific and engineering enterprise and its educational activities in ways other than those encompassed by the first three criteria. Included under this criterion are questions relating to women, minorities, and the handicapped; the distribution of resources with respect to institutions and geographical area; stimulation of quality activities in important but underdeveloped fields; and the utilization of interdisciplinary approaches to research in appropriate areas.

The merit selection criteria have evolved over the years. At one time there were separate criteria for large-scale facilities and centers (see Box 3-1).

Review and Selection Criteria for Major Project Awards

NSF procedures permit special treatment of major award projects. Criteria in addition to the four basic ones may be applied to specific programs. They are supposed to be listed in program announcements and solicitations; specified in the cover letters sent to peer reviewers; and used by site visitors, panels, and NSF staff in summarizing the basis for their recommendations. In addition, NSF still has "special guidelines for organization and research administration" for large projects, which were based on the 1974 criteria for creating or changing institutional structures:

    a. *Criterion of Need*: If a new administrative structure is proposed, evidence should be provided that it is needed to address scientific problems in a manner or on a scale not possible with existing structures.

    b. *Criterion of Long-Range Potential*: Evidence should be provided of a mission with such potential for high scientific productivity over an extended time period, that a significant number of excellent scientists are willing to commit their careers to it.

These criteria, which are descended from the longer list of criteria adopted by NSB in 1967, are not now part of any formal planning or review procedures although they are in the *Proposal and Award Manual* (NSF, 1993c:I-4). They were not used in the 10 case studies, or at least they are not mentioned in the documentation (although the criteria in some cases, for example, for Engineering Research Centers (ERCs) and Science and Technology Centers

BOX 3-1: Evolution of Merit Review Criteria

The criteria used to review research proposals have been elaborated and added to over the years, but the primary emphasis has remained on the promise of the proposed research and the competence of the researcher. The first formal statement of selection criteria was approved by NSB in 1967. That statement reaffirmed NSF's practice of managing its basic research program "in such a way as to permit the development of science along lines dictated by the internal needs of science itself" (NSB, 1967:3). To achieve this goal, NSF looked to fund unsolicited proposals from university-based researchers who met certain criteria: promise of significant scientific results; past record of performance of the investigators; potential scientific impact of the proposed work; degree of novelty, originality, or uniqueness involved; educational value of the proposed research in terms of effects on students; and relevance of the proposed work to potential applications.

NSB also recognized the place of national facilities and noneducational research institutions in a "balanced" science program. Examples at that time included the "national research centers" for radio and optical astronomy and atmospheric research, with their very large and expensive facilities and equipment, and large-scale research programs requiring logistical support and coordination among otherwise independent researchers. To determine how to support such large-scale activities, NSF had to make three types of decisions: whether to create a new research institution, augment a facility already in existence, or phase out or transfer programs from a facility that was no longer appropriate. NSB (1967:6) adopted a separate set of criteria for these decisions (although they are more relevant to deciding whether or not to initiate a major project than to deciding among competing proposals).

1. Does the laboratory meet a real scientific need and an opportunity to attack important problems in a way, or on a scale, not otherwise feasible or promising? Is there a broad mission which is sufficiently specific to offer a continuing challenge to the laboratory with consequent assurance of high scientific productivity over an extended period? Have the requirements for continued evolution of capabilities and facilities been given adequate consideration in preliminary planning?

2. Is there and will there continue to be a significant number of first-class scientists (as judged by their peers) who believe deeply in the proposed program and are willing to stake their personal scientific

Box 3-1 continued

reputations on its success, including direct involvement in the program on both a full-time and a long-term basis?

3. Are there convincing arguments that the program objectives can better be achieved through the organization of a new program at a national center than through existing academic or other research institutions? To what degree would the new capability under consideration be unique on a national basis?

4. Will the center or its programs strengthen or detract from related work in the universities? Will the center provide new research opportunities for academic and other scientists? Is there assurance that user scientists will be accepted into the facility primarily on the basis of the scientific merit of their projects?

5. What contributions will the work of the laboratory make to the training of future scientists and/or technologists, including the training of future potential faculty members and industrial investigators as well as students generally? Will the laboratory foster transfer of new basic research techniques into technology and into other areas of science?

6. What impact is the work of the laboratory likely to have on other areas of science?

7. To what degree may tangible social benefits ultimately emerge from the work of the laboratory? The ultimate social benefits of the work of the laboratory are extremely difficult to foresee; hence, significant fundamental research programs should not be rejected because of inability to apply this criterion in a meaningful manner. By the same token, proposals for major programs, which argue their cause on the basis of intrinsically dubious forecasts of social benefits, require the most careful evaluation.

In 1974, NSB revised and elaborated the selection criteria it had developed for individual investigator proposals and made them applicable to all proposals, from small research projects to large-scale national facilities. The criteria were grouped in four categories relating to competent performance, intrinsic scientific merit, utility or relevance, and future scientific potential of the nation.

> Box 3-1 continued
>
> Only a subset of the planning-type criteria developed in 1967 for the national research facilities and large-scale coordinated research programs was carried forward in the 1974 formulation of the criteria to be used for proposals to "create or modify institutional and organizational structures." They were criteria of need (evidence of real scientific need; an opportunity to address important problems in a way or on a scale not otherwise possible) and criteria of long-range potential (well-defined mission with prospects of high scientific productivity over an extended period; evidence that a significant number of first-class scientists will stake their careers on it; evidence that the new organizational entity will strengthen rather than displace related work in educational institutions).
>
> The current review criteria for all proposals were approved by the NSB in 1981. The four basic criteria (described in the text) correspond to the four categories for individual and small-group research projects first identified in the 1974 guidelines. Provision was also made for additional specific criteria to be listed in the program announcement or solicitation for a specific project (NSF, 1992a:I-3).

(STCs), did cite the advantages of the center type of organization).

Major awards almost always involve criteria in addition to the basic four. They are listed in the solicitation announcements (see Appendix D for recent examples of selection criteria).[5] The additional criteria can make decisionmaking more complex and difficult. The criteria may include, for example, managerial capacity, technology transfer, human resource development, participation of underrepre-

---

[5] Additional criteria are not necessarily confined to large facilities and centers that are the subject of major awards. For example, proposals for materials research groups (MRGs) are evaluated according to the four basic criteria plus two key features of an MRG (NSF, 1992b): (5) need for MRG funding mode ("an assessment of the need for a group approach in order to make significant advances, or evidence that advances in the topic cannot be made via the individual investigator approach") and (6) degree of interconnection ("an assessment that the research plan constitutes an integrated and collaborative effort").

sented groups, and/or nonfederal cost sharing. The relative weighting of these criteria is rarely specified in the solicitation document and is indicated only generally in the letter to reviewers. The problem of priority weighting is further complicated by uncertainty about when in the review process each of the various criteria should be addressed. It also may not be clear which reviewers were selected to apply which criteria, making it difficult to know how to weigh the review comments.

Findings and Recommendations on Criteria

The panel found that for major awards the review criteria and their relative importance were not always well understood by participants, including proposers and peer reviewers, or by knowledgeable observers. Peer reviewers often confine themselves to those criteria they are most qualified to judge, usually technical merit and capability of the proposer, and it is not always clear who is assessing the other criteria or how they are eventually integrated by staff in the final decision. The panel concluded that NSF should (1) state carefully and clearly the criteria that will be used in the review process, and (2) be explicit about the relative importance of the review criteria.

The panel does not advocate the use of strict numerical weightings, which is the approach taken, for example, in procurement contracts. Choosing the best research-related proposal necessarily involves a large degree of expert judgment and some degree of uncertainty; thus, decisionmaking about awards in these cases cannot be reduced to a quantitative algorithm. Nevertheless, without any guidance at all, reviewers are implicitly allowed to derive their own weights for the various criteria when arriving at their overall rating; NSF officials and NSB members also must use their own judgment in evaluating proposals against the criteria. We believe that it is possible and desirable to describe the relative importance of the criteria specified for a major award project to help ensure that reviewers give highest consideration to proposals best able to meet the research goals of the program.

Recommendation 3: Primacy of Technical Merit Criteria

The NSB and NSF should continue to make technical excellence the primary criterion in evaluating the merit of proposals for major awards. To ensure that research funding is used most effectively, no major award should ever be made to a project that is not of very high technical merit. Additional criteria should be used only to choose the best overall proposal from among those whose technical merit is among the most highly rated.

Research results are the main goal of a major award, although major awards may also promote the realization of other meritorious goals. Technical merit is the likelihood that a proposed project will achieve the projected research objectives, whatever they may be. Thus, technical merit has been and should remain the primary standard met by all NSF awards, including major awards, because it best assures effective use of the funding devoted to federally supported research. It is especially important in major awards because there is less opportunity for trial and error than in a program of small awards. Emphasizing technical excellence also gives merit review added credibility that should help forestall academic earmarking or other political interference in award decisionmaking at NSF.

For smaller awards to individual investigators, technical merit criteria have to do with the originality of the project, soundness of the methodology, and qualifications of the researcher to carry out the proposed research. In major awards, technical merit is more complex because the factors involved in successful implementation of a large-scale, long-term project are more extensive. For example, these factors may include such considerations as the managerial capacity of the principal investigator or investigators, and their institution or institutions, to implement and operate a major research facility or center; physical site characteristics if construction is involved; or the involvement of industry and technology transfer plans if an applied research effort is involved.

Major awards usually involve criteria in addition to the technical merit of the projects and the capacity of the proposers to carry them out. These additional merit criteria include (but are not limited to) human resource development potential, participation of women and minorities, geographical and institutional distribution and capacity building, and/or relevance to national socioeconomic goals. Peer reviewers selected for their technical expertise may not feel qualified to apply the nontechnical criteria, although they may comment on them. Because the additional considerations must be taken fully into account by NSF officials in recommending the best overall proposal, reviewers should be appointed specifically to assess each of them and provide input to the process.

This recommendation would be more easily implemented in conjunction with the two-phase review process recommended later in this report, which provides a mechanism for systematically ensuring that both high technical merit and the other criteria are taken into account in reaching a final decision on the overall merit of major award proposals.

The precise dividing line between technical and nontechnical merit criteria cannot always be determined a priori; it is a decision contingent on the nature of each project's objectives. The definition of technical criteria must be determined as part of the design of the review process for each project (recommended later).

The technical merit criteria should measure a proposed project's likelihood of achieving its primary scientific or engineering research objectives. These would be different for different types of projects. For example, a proposal to construct and operate a national research facility with state-of-the-art instrumentation would have to show not only that top scientists were involved and first-rate research was likely in the in-house program, but also that the engineering requirements and management capacity made the project feasible. A proposal for an interdisciplinary research center would have to show promise of "centerness," that is, that the center would be organized and managed in such a way that the research done would contribute more than an equally expensive program of individual grants.

In each case, what is most important is that a clear decision be made about which criteria will be considered technical (e.g., integral to the project's research achievements) and which are secondary (e.g., desirable in addition to research results). Other criteria may be ones that qualify proposals for consideration in the first place (e.g., a cost-sharing requirement or, in the case of a facility, the absence of physical problems such as seismic activity or light pollution). The criteria, their classification as technical or nontechnical, and their relative importance should be stated clearly in the solicitation announcement.

Recommendation 4:   Human Resource Development and Equal Opportunity as a Criterion

The contribution of every major award proposal to overall human resource development should be emphasized. The number of students to be involved—and the inclusion of minorities and women at all levels, from students to senior investigators and project managers—are important components of human resource development and equal opportunity. They should receive more explicit attention in the review process.

Research opportunities for minorities, women, and other underrepresented groups constitute a criterion second only to scientific and technical merit. Long-term scientific progress depends not only on today's research but also on tomorrow's well-trained and experienced researchers and research leaders. The research enterprise is also a major source of trained personnel for industry—an important mechanism for the transfer of knowledge from research to product development. Any major investment of NSF's resources should be assessed for its potential impact on the training and education of scientists and engineers, especially minorities and women who are currently underrepresented in the nation's technical work force.

At the same time, the women and minority investigators already in science are underrepresented in research leadership positions, particularly in the major awards category. In the case studies it examined, the panel did not find a single female principal investigator, and it seems likely that minority investigators were equally rare. Given the gradual shift in NSF resources toward this category of award, special attention should be devoted to opening the door of opportunity for leadership on major award projects at least as wide as it has been opened for women and minorities to be principal investigators on individual investigator grants.

Accordingly, all proposals for major awards should document the expected number of students who will be involved and the plans for participation of minorities and women at all levels. Once the primary criterion of technical quality is met, proposals that emphasize human resource development and equal opportunity features should be given strong consideration.

Recommendation 5: Cost Sharing as a Criterion

Cost sharing should be used only to demonstrate commitment to the project's goals and never simply to extend NSF funds. Where cost sharing is required, NSF should spell out its expectations in the solicitation announcement. The amount of credit for cost sharing for purposes of evaluating proposals should be stated clearly and capped at a reasonable level. Due account should be taken of the likelihood that cost-sharing commitments will in fact be met in the out years.

The level of nonfederal cost sharing in major awards has been increasing for a decade. In some cases, the lack of clarity about the matching requirement has caused problems in the review process.

Nonfederal cost sharing may be a legitimate criterion for a major award, but in accord with NSB policy, it should be required only for programmatic reasons (e.g., to ensure industrial relevance of the proposed work, to strengthen its linkages to an interdisciplinary research center, or to show that a university is committed to running a high-quality national facility). Open-ended cost-sharing requirements may place an unintended and unfair burden on universities and state governments, and interfere with their ability to order their own research support priorities.

In some cases, cost-sharing requirements may not be achievable by potential proposers in states where legislatures are not in session, for example, because they only meet biennially. Most importantly, excessive cost-sharing requirements can severely limit the scope and quality of the proposals received.

Concerned with the rapid increase in cost sharing, an NSB committee studied the issue in 1986 (NSB, 1986). The committee's conclusion was adopted by the NSB and reaffirmed after a staff update of the study in 1989 (NSB, 1989):

> The Committee finds that the Foundation has used leveraging in the form of matching requirements in funds, people, and equipment primarily to achieve substantive objectives rather than merely stretch its budget to do more, and recommends that this emphasis be continued.

The panel found that unclear cost-sharing requirements had caused problems in the review of several major awards. In the case of the Earthquake Engineering Research Center, for example, a 50-percent match requirement deterred some potential proposers from applying because they could not obtain $5 million a year from their states (GAO, 1987:31). The lack of clarity about the nature, source, and timing of the match also caused problems for several that did apply. The match for one proposal came very late in the process because the state legislature had not met, and NSF "conditionally" recommended that the award be made to another proposal before the review of the first had been completed (GAO, 1987:32).

In the case of the National High Magnetic Field Laboratory, the project development plan included a 50-percent matching requirement. When the solicitation was drafted later, it said only that the facility would be "heavily cost-shared" (NSF, 1989). The winning proposal was able to obtain $58 million over five years in matching funds from the state government that, with some university funds, was a little more than the 50 percent match called for in the project development plan. This substantial sum was more than any other applicant could secure and appeared to many in the research community to have an undue influence on the award decision.

Although the panel's recommendation is current NSB policy, it is reaffirmed here because the temptation to leverage other funds to make NSF dollars go further is very strong. The level of nonfederal cost sharing in NSF awards has been increasing over the past decade. That increase occurred in part to ensure local commitment or the relevance of interdisciplinary research to industry, but there comes a point at which the burden on local institutions causes local problems. Accordingly, NSF and NSB must ensure that the cost sharing is justified on a case-by-case basis and does not impose an excessive net burden on universities and states, or unduly distort their priorities or the priorities of NSF. They also should take into account the sources and quality of the cost-sharing commitment as well as its size.

NSF PROCEDURES FOR REVIEWING PROPOSALS

The review part of the award process usually commences when NSF releases the solicitation document. (Some major awards are not competitively solicited, and the review begins when the proposal is received.) The review phase culminates in NSB's review and approval of the award. It includes the peer review of proposals, the decision by staff to recommend funding of a particular proposal or proposals, endorsement by the NSF director, and review and ratification (or disapproval) by NSB.

Proposal Review Process

Once the criteria are developed for choosing among project proposals, NSF must design a process to ensure that the criteria are applied consistently and appropriately. In a merit review system, this means that individuals with the expertise to assess each criterion must be appointed and deployed as needed for that specific solicitation (e.g., ad hoc reviews by individual experts, perhaps site visits, and one or more panel reviews). Finally, these judgments must be weighed and combined in an overall ranking of proposals for decisionmaking at higher levels within NSF. In major award cases, the NSF director recommends a decision and presents it to NSB for approval. As noted above, technical excellence in carrying out the research-related objectives of the project should be paramount in choosing among proposals, but other criteria may and usually should be taken into account in choosing among top-rated proposals.

CRITERIA AND REVIEW PROCEDURES

Peer Review Modes

The NSF program director decides on a review process for each proposal that will enable him or her to make a sound, well-documented recommendation. Several "tools of the trade" are in common use at NSF, often in combination. They must include some form of external peer review, such as the following:

1. *Ad Hoc Mail Review*: The NSF program director sends the proposal to at least three (usually four to eight) experts on specific aspects of each proposal, who are asked to submit written reviews to the program officer.

2. *Panel Review*: In addition to or in place of mail review, a panel of experts may be convened to evaluate and rank proposals. In some programs, especially biology and earth sciences, standing advisory committees meet two or three times a year to consider proposals. In other programs, such as ocean sciences, ad hoc panels are formed to review each round of proposals.

3. *Site Visit*: For large projects, competitive solicitations for centers and facilities, or particularly difficult or unusual proposals, mail and panel reviews may be supplemented with site visits by teams of outside experts and NSF staff. NSB members may be site visitors as well. Site visits are generally also used for decisions on renewing projects.

The program director evaluates each proposal and may make a site visit, invite reviews by colleagues at NSF with appropriate expertise, or consult with other federal agency officials.

Selection of Reviewers

Program directors are required to select a set of reviewers that, in the aggregate, can provide the information needed to make a recommendation in accordance with the selection criteria.

Selection of Mail Reviewers. The virtue of mail review is that each proposal may be reviewed by the experts most knowledgeable about the subfield involved. Ideally, reviewers as a group will be able to evaluate a proposal with respect to all of the selection criteria. Therefore, the balance desired among criteria should affect the selection of reviewers. In selecting reviewers, program directors are also asked to take into account any additional criteria stated in program announcements and solicitations.

The award manual lists the "optimum" criteria for selecting reviewers (NSF, 1993c:I-5). For all proposals, some reviewers should be experts in the research subfields involved so they can evaluate the proposals on the criteria of competence, intrinsic merit, and utility. If the proposals involve substantial size or complexity, broad disciplinary or multidisciplinary content, or potential applications to significant national problems, some reviewers should have a broader or more generalized knowledge of the research subfields involved. Some reviewers are supposed to have a broad knowledge of the infrastructure of the national science and engineering enterprise to evaluate proposals for their impact on scientific and engineering education and human resource development, distribution of resources to institutions and geographical areas, and other social goals. Finally, reviewers as a group should reflect a balance among various characteristics such as geography, type of institution, and underrepresented groups.

Selection of Panel Members. The advantage of panels is the opportunity they provide for face-to-face interaction among reviewers and between reviewers and the program director. This interaction allows a more detailed discussion of each proposal and how it relates to the overall program. On the other hand, a relatively small group reviewing a variety of proposals is less likely to be as knowledgeable about every subfield involved as mail reviewers can. That is especially true of a standing advisory group. The NSF award manual notes that "it is important, nevertheless, that such groups be structured to provide broad representation and many views on matters under the advisory group's purview" (NSF, 1993c:I-6). The manual offers some general considerations that should be taken into account in achieving balance in advisory groups. They include

- individual qualifications;
- fields of expertise;
- public impact—some panel members should represent regions, organizations, or segments of the public directly affected by issues under consideration;
- academic and nonacademic impact—members should represent small, medium, and large institutions, as well as public and private institutions; representatives from outside the academic community are also desirable in most instances;
- underrepresented groups;
- age distribution; and
- geographic balance.

Careful selection of a balanced set of reviewers is especially important for major awards, not only because they are more visible but also because they are more complex. It is critical that the reviewers are—and are perceived to be—qualified, unbiased, and balanced as a group. NSF was criticized in the case of the Earthquake Engineering Research Center, for example, for having too few earthquake researchers and too few Westerners on the review panel (GAO, 1987).[6]

---

[6] The General Accounting Office (GAO) audit report on the decision concluded that the composition of the panel had met NSF's requirement that there be reviewers with both special and general knowledge of the scientific subfields involved in the proposal. Although only one of the seven panelists was a recognized earthquake engineer, three others had some previous experience in earthquake engineering. Most were chosen for their experience in managing large research efforts at a university or in industry. GAO also had the technical sections of the leading proposals reviewed by four nationally known earthquake engineering experts; according to GAO, their assessments of each proposal's strengths and weaknesses concurred with those of the NSF-appointed panel (although their comments were more numerous and more detailed) (GAO, 1987:20).

Policies and Procedures for Dealing with Bias and Conflict of Interest

Merit review at NSF has been criticized many times as an "old-boy network" that concentrates awards in a relatively small number of institutions in a few states. Various policies and procedures have been added over the years to ensure that the system is open and fair. NSF program directors are not supposed to use anyone as a reviewer who (1) would be directly involved in the project as a collaborator or consultant, (2) is from the same institution as the applicant, or (3) has a family relationship with the applicant.

The program director is also required to document the existence of interests, affiliations, and relationships that might affect a reviewer's evaluation of a proposal. Reviewers are explicitly asked to describe relationships that may be or appear to be a conflict of interest. The program officer is required to indicate in the file how such conflicts or potential biases were handled in making the award decision.[7] Reviews that contain intemperate personal attacks or other indications of bias are not supposed to be used in the selection process. Other policies and procedures to ensure fairness are described in Box 3-2.

Award Decisionmaking

Among the key actors in the decision process are NSF program directors. They make initial recommendations for the review format and specific reviewers to be used for each proposal or competitive set of proposals, which division directors and assistant directors review and approve. The program directors recommend whether or not to fund a proposal, as well as the amount and duration of support. They base the recommendation on the comments of reviewers, but they

---

[7] NSF also has special procedures for dealing with proposals from individuals who are prospective, current, or former employees, NSB members, or Intergovernmental Personnel Act assignees at NSF.

Box 3-2: Other NSF Policies and Procedures to
Ensure Fairness in Merit Review

Since the mid-1970s, NSF has taken a series of steps to open up the entire award decisionmaking process, including peer review. The intent of these changes was spelled out in 1977 (NSB, 1977:iii):

> . . . the peer review process should continue to preserve the traditional benefits of peer evaluation of intrinsic scientific merit. At the same time, it is essential that the research community perceive the peer review process to be fair, and equitable as well as accessible to all qualified persons—both as research applicants and as reviewers.

Preserving the traditional benefits of peer review has meant maintaining the anonymity of reviewers, because NSF believes reviewers will not be candid if applicants know who they are. NSF has, however, acted to achieve the following:

1. Provide applicants with (a) verbatim mail review remarks without identifying the reviewers, (b) the panel summary if a panel was used, and (c) site visit reports if a site visit was made.[1]

2. Let applicants suggest the names of individuals who should not review their proposal.

3. Give every applicant a description of the contextual factors considered in making a decision and, in the case of declinations, an explanation of the basis for the decision (since 1990). Contextual factors are those beyond reviewer comments and ratings that staff take into account when making the award decision. Such factors include the total amount of funds available to the program for new and renewal proposals, the number of proposals the program expects to review that year and the approximate percentage that will be funded, the need to maintain balance among research subfields, and the availability of funding from other agencies.

4. Allow the principal investigator (PI) of a declined proposal to ask the program officer or division director for additional information; if the PI still believes the decision was not fair or reasonable, he or she may write to the assistant director requesting a formal reconsideration. If the declination is upheld, the PI's institution may appeal to the NSF's deputy

Box 3-2 continued

director for a second reconsideration.[2]

5. Maintain ongoing monitoring and analyses of the peer review system. The director of NSF has submitted an annual report on peer review to NSB since 1977, with basic statistics on the number of reviews, approval rates, and appeals, as well as a listing of special waivers from peer review. The NSF Office of Inspector General audits a random sample of proposal jackets each year to see how well NSF is following its own policies and procedures. External peer reviews of program operations by Committees of Visitors evaluate each program every three years by reviewing a sample of proposal jackets.

---

[1] The actual peer reviews have been included in the proposed award packages going to NSB since March 1976 (Windus, 1984:17).

[2] In 1990, the grounds for reconsideration were expanded from procedural due process to include substance (NSF, 1990d). Requests for reconsiderations are few, therefore, and reversals are rare. In FY 1991, for example, there were 39 first-level and three second-level reconsideration actions. NSF's original decision was upheld in all but one case (Massey, 1992). Most PIs prefer to revise their proposals or submit them somewhere else rather than appeal.

---

may take other considerations into account, such as the balance among research topics or the amount available in the program budget.

In the case of individual investigator grants, recommendations by program directors can be approved at the division level. If the proposal is large enough, as in the case of major awards, the award decision is reviewed by the appropriate assistant director and at the NSF director's level before being sent for approval to NSB with a memorandum from the director.

Although external reviewers are asked to evaluate a proposal's impact on geographic and institutional balance and on participation of women and minorities, the program director and other NSF staff are often the only ones in a position to make these judgments in the case of a program of small grants. Facing proposals of nearly equal merit, the program director may give greater consideration to scientific program balance and distributional factors (geographic, institutional, race, sex, etc.). This is especially true in small grant programs as they near the end of their available funds, when program directors may achieve greater balance among secondary criteria by selectively choosing slightly lower-rated proposals.[8]

In major award cases, however, secondary criteria cannot be balanced by letting NSF staff select some lower-rated proposals at the margin, after funding the best proposals technically. For major projects, only a few awards, or often just one, is made. This means that all the criteria, even though they are more numerous and complex than for small grants, must be met in one or a few proposals. Given the importance and complexity of major awards, the role of outside peer evaluation in weighing and balancing the various criteria, both technical and nontechnical, is especially important throughout the process. The recommendations in this report thus aim to create a merit review system in which outside peer review not only plays its usual major role in evaluating technical excellence but also plays the major role in evaluating secondary criteria and identifying the most meritorious proposal based on all criteria.

Findings and Recommendations on Review Procedures

The panel found that the review processes for major awards vary greatly. Some involve elaborate and well-thought-out procedures for applying different types of criteria and expertise at different levels

---

[8] In practice, proposals are rarely put in strict rank order because the proposal evaluation process is too imprecise; rather they are informally grouped. Even so, NSF does not usually have the budget to fund all the proposals in the top or excellent group. According to NSF officials, that means that using additional criteria to choose among them at the margin does not lower the average quality of the research supported.

and stages of the review. Some are simply handled as investigator-initiated proposals.

At one time, NSB had criteria and procedures for large-scale facilities and centers that were different from those for the small-project grants. Over time, however, the procedures for the latter came to be applied to all awards, regardless of their nature, size, or complexity. *The panel concluded that the major awards are different enough from the typical small grant, and alike enough because of their size and impact, to warrant standard procedures of their own.* The key is to structure the review process so that the various types of criteria are applied by appropriately qualified reviewers.

Recommendation 6: A Two-Phase Merit Review Process

For major awards, the peer review part of the merit review process should be conducted in two phases. The first phase would be a strictly technical review; to help assure the primacy of technical merit, only those proposals judged to be technically superior would be forwarded to the second phase for any further consideration. In the second phase, the additional merit criteria would be weighed and balanced with the technical criteria by a more broadly constituted group of reviewers. This second-phase panel would recommend the proposal (or proposals) best meeting the full set of criteria. If the proposal judged to have the highest merit overall is not the one ranked highest in the first phase of review for technical merit, the second-phase panel must explain its recommendation fully. If the top-ranked technical proposal is subsequently not recommended by NSF staff, the chair of the first-phase panel or another member of that panel should present the case for it at the NSB level.

This two-phase methodology, which NSF and NSB have already employed successfully for some major award competitions, will help ensure that three important goals of the review process are met.

CRITERIA AND REVIEW PROCEDURES   85

First, NSB will approve awards only to projects that meet the highest technical criteria and thus have the greatest promise of achieving their research-related objectives. At the same time, however, it will ensure that other important but secondary merit criteria are evaluated by qualified reviewers and appropriately weighed in the final ranking. Third, the process will better clarify the basis for the final decision and the respective input of peer reviewers and agency staff, which in turn will make the decisionmaking process more acceptable and accountable to both the research community and the broader public to which NSF must answer.

Procedures. If a large number of proposals is expected, NSF should use preproposals to narrow the field. The technical merit of the preproposals should be peer reviewed, either through the use of mail reviews or perhaps through a meeting of a Phase 1 review panel.

Phase 1 is the technical review by relevant experts. It can be done in stages or steps, including mail reviews, site visits, and panels or subpanels. In the final step of Phase 1, a panel of technically qualified reviewers evaluates all the proposals for technical merit (with input from the mail reviews, site visits, etc.), sorting them into no more than four or five categories of excellence, ranging from most qualified to unqualified. The panel should strive to produce a priority ranking of the proposals if at all possible. If the ranking by quality is reasonably clear, the panel should forward the proposals in rank order. If there are a number of excellent proposals, each with its strengths and weaknesses, the panel may forward a group of top-ranked proposals judged to be essentially equivalent in technical quality. In this case, the panel should comment carefully on the strengths and weaknesses of each proposal.

In research center competitions where there are numerous proposals, it is relatively easy to forward a set of top-rated proposals to the second phase, explicitly ranked in order of technical merit. In fact, NSF already uses a form of the two-phase merit review process for the Engineering Research Centers, Science and Technology Centers, and Supercomputer Centers. The more difficult cases are competitions in which there are only a few proposals. We expect on the basis of the case studies, however, that there will usually, if not always, be at least several strong candidates in the second phase of a major award competition.

Phase 2 is a review of all other merit criteria (e.g., human resource development and equal opportunity, geographic or institutional distribution, technological relevance, industrial involvement, outreach and technology transfer activities). To complete this phase, the subgroup of proposals ranked "excellent" or "most qualified" in Phase 1 panel is considered by a new, Phase 2 panel of reviewers. This new panel should be qualified to evaluate proposals on both the additional criteria and the technical criteria identified for that project (although it should not redo the technical ranking). Additional review steps could occur during this phase, such as special subpanels to consider certain criteria (e.g., business people and economists to evaluate proposals for their potential contributions to economic competitiveness), additional site visits, or presentations by finalists.

The Phase 2 panel would recommend the proposal (proposals if there are multiple awards) that best meet the full set of criteria. If the recommendation is for a proposal that was not the most highly ranked technically in Phase 1, the Phase 2 panel would have to explain explicitly the reasons for its action. In that case, or if NSF staff decides to recommend a proposal to NSB that was not ranked highest by the Phase 1 review panel, NSB or at least its Committee on Programs and Plans (CPP) should hear a presentation in favor of the top-ranked technical proposal from the Phase 1 panel chair or another advocate from that panel. He or she should present to CPP the case for the best Phase 1 proposal and help CPP understand the research-related benefits that might be forgone if another proposal were chosen. Currently, CPP-NSB hears only the case in support of the proposal recommended by NSF and thus cannot make a decision on the basis of a comparative analysis of the top few proposals. We believe that NSB would be in a better position to make a more informed decision if it more clearly understood the trade-offs among merit criteria involved in choosing the winning proposal and that this understanding is particularly important when the successful proposal is not the one with the highest technical merit according to the external peer reviewers.

Reviewer Roles. Reviewers in Phase 1 should be chosen so that, as a group they will be able to evaluate and rank the proposals according to technical merit (as measured by the technical criteria specified in the solicitation). In addition to providing an overall summary rating based on technical criteria, Phase 1 reviewers should be encouraged to comment, if they wish, on the other merit review criteria. In this case, they should give two ratings, one based on purely technical criteria and the other indicating how they might change their overall ratings in light of additional nontechnical criteria (at the same time, indicating their competence to apply the other criteria).

So far, the process is not very different from that employed in many cases, except that the procedure recommended here calls for the use of a panel in every case to integrate the reviews rather than rely on mail reviewers and staff alone. The panel is also charged with deriving an overall ranking of proposals according to technical merit only, without taking other criteria into account.

A new panel is required in Phase 2, although mail reviewers, subpanels, and site visitors may also be used. The Phase 2 panel should be constituted to apply the full set of merit criteria, technical and nontechnical, because this group has the task of identifying the best overall proposal. It should be noted that the eventual winning proposal has already had to achieve a high threshold of technical excellence; thus, no award would go to a project of low technical quality just because it ranked high on other merits. A winning proposal must rank high in both areas, technical and nontechnical.

At least some Phase 2 reviewers should be carried over from the first-phase panel. This continuity is helpful because the primary and secondary criteria usually interact in ways that need to be evaluated before reaching a final ranking. For example, one of two proposals basically equal in scientific promise might better involve students because of the way the research is organized or where it is done. That would affect the rating positively when education is an important secondary criterion.

Phase 2 panels should be asked to provide a careful qualitative discussion of how they balanced the nontechnical criteria against the technical ones. We believe that in the typical case the discussion will revolve around whether proposals ranked second or third on the technical merit scale are the most meritorious overall and should be funded. In some cases, several proposals may have substantially equal technical merit. In these cases, important secondary criteria may play a tiebreaker role. Even here, however, each proposal will have its individual technical strengths and weaknesses, and it will be helpful to have carryover members from the Phase 1 panel familiar with the proposals as participants in the discussion of the balancing of technical and other merits. As already mentioned in Recommendation 3, if the Phase 2 panel decides to bypass the proposal or proposals ranked highest technically, it should provide a careful written explanation of the justification for doing so.

4 Awarding Major Projects: NSB Role, Review Process Design, and Decision Documentation

This chapter addresses the three final aspects of the proposal review and award phase for major projects: the role and procedures of the National Science Board (NSB), the requirements for designing the review process and issuing the solicitation announcement, and documentation of the basis for the final decision.

NSB ROLE AND PROCEDURES

If awards are very large (at least $6 million over five years or $1.5 million a year) or if they pose new programmatic or policy issues, they are subject to review and approval by the NSB. After a proposal is recommended for funding by the program director, an NSB decision package is prepared and put through a three-stage internal review process that begins about six to seven weeks before the NSB meeting (NSF, 1993b). First, a directorate review board reviews the NSB package for merit, adequacy of merit review, and completeness. Second, a new body—the Administrative Review Group (ARG)—reviews the NSB package for compliance with the administrative policies and procedures of the National Science Foundation (NSF). ARG members include the executive secretaries of the NSB committees on Programs and Plans (CPP) and on Education and Human Resources (EHR), and the NSB executive officer. They also include representatives of the Office of General Counsel (OGC), the Division of Grants and Agreements, and the Budget Division.

After ARG review and sign-off, the assistant director for the originating directorate approves the award and forwards the NSB package to a third-level review by the Director's Action Review Board (DARB) before final sign-off by the director about two weeks before the NSB meeting. DARB conducts a final review of the proposed award on behalf of the director and discusses issues likely to be raised by the NSB. DARB is chaired by the deputy director and includes several assistant directors and other staff chosen on an ad hoc basis by the deputy director.

DARB plays a very important gatekeeper role in the review process. It stands between the directorate that wants to make an award and the director. DARB recommends to the director whether or not (1) the review plan and timetable for each major project solicitation are adequate and (2) the case for the award decision has been made and documented adequately in the materials forwarded to NSB. If DARB asks hard questions about procedural and substantive issues, the burden on NSB to review every award decision is lessened, which allows it to devote more attention to long-range planning and program balance issues.

After it has cleared DARB, NSB receives the decision package, which includes

- a director's memorandum that summarizes information and issues related to the proposed award;
- the program officer's recommendation;
- the summary budget;
- a list of reviewers and review analysis (Form 7); and
- verbatim peer reviews.

The director's memorandum, which is prepared by the originating program staff, is supposed to include "objectives, alternatives, potential policy implications, precedents involved, and any other factors that could be considered nonroutine" (NSF, 1992c:III-2). It should also include a short statement of responses to the major

NSB ROLE AND REVIEW PROCESS DESIGN 91

concerns raised by reviewers and an analysis of immediate and long-term budget implications.

Other awards "of unusual sensitivity" must must be submitted to the NSB for approval. Those involve

- special policy issues;
- the establishment of new centers, institutes, or facilities;
- potential for rapid growth in funding or special budgetary initiatives;
- research community or political sensitivity;
- a previous expression of NSB concern; and
- any other awards selected by the director or assistant directors.

NSB conducts its work through committees. Program planning and major award reviews for research and related activities are assigned to CPP (the same functions for education activities are performed by EHR). NSB met an average of seven times a year in 1990-1992, up from six a year in 1986-1989. CPP meets several times to review recommended awards during each NSB meeting, which usually last two or three days. The June meetings of NSB are reserved for long-range planning, although the heavy CPP review schedule required it to approve specific awards during the June 1992 planning meeting (Appendix B lists the awards approved at each meeting from 1986 to the present).

CPP also reviews and recommends approval to the full NSB of all proposals to start new research programs (e.g., a program of arctic social science research was approved in October 1989), project development plans for proposed new or revised projects (e.g., National Nanofabrication Users Facility in November 1992), and proposed solicitation announcements.

CPP, and NSB as a whole, must divide their attention among long-range planning, broad program oversight, and specific award approvals. The number of awards subject to NSB approval has been increasing in recent years, causing workload problems.

In FY 1992, for example, 49 of the approximately 9,000 awards made by NSF went to NSB for approval, totaling nearly $1 billion (up from 37 for $520 million in FY 1991 and 28 for $530 million in FY 1990). Of the 49, 31 were for research centers, facilities, and other large research projects (the other 18 were education projects).[1]

As the number of projects coming to it has increased in the last several years, NSB has become concerned about having to spend most of its time reacting to specific project awards. It has discussed various ways to reduce the workload, including

- raising the delegation thresholds of $1.5 million a year or $6 million total;
- concentrating on approving research center _programs_ rather than individual _project_ awards made under the programs, or approving the first round of individual center awards but subsequently reviewing only the program as a whole (letting the staff decide on renewal awards after full merit review); and
- greater use of the authority to waive reviews for routine types of decisions, as is already done with renewals of materials research laboratories and some smaller astronomy and physics facilities (again, letting the staff decide on renewal awards after full peer review).

The NSB has endorsed the last option and has begun to make greater use of waivers during the past year.

---

[1] The 31 research-related projects reviewed by the CPP and approved by the NSB included 11 Science and Technology Centers, nine Engineering Research Centers, four Materials Research Laboratories, the Ocean Drilling Program, the National Center for Atmospheric Research, the Synchrotron Radiation Center at the University of Wisconsin, the National Superconducting Cyclotron at Michigan State University, and several other facilities and projects (see Appendix B).

Findings and Recommendations on the NSB Role

<u>Recommendation 7: Reorienting the NSB Workload</u>

NSB should manage its proposal review workload to ensure that adequate time is left for its most important activities of broad policy direction, long-range planning, and program oversight. That could be accomplished by using its exemption authority more frequently, or by raising the delegation threshold, or both.

The panel found that CPP spends much of its time responding to recommendations for specific awards. That leaves less time for it to examine overall program direction and balance or to assess the continued justification for major facilities and programs. Moreover, the number of proposals that CPP and NSB must review for awards is increasing. The panel concluded that this trend is impinging on NSB's most important functions—broad oversight and policy direction.

NSB should devote its energies first to planning (see Recommendation 1), second to overseeing the design of solicitations, and third to reviewing specific decisions on very large or significant awards.

The greatest increase in workload has come from the creation and expansion of the centers programs in the mid-1980s, in which individual centers receive full merit reviews every three years. The NSB could concentrate on evaluating the centers as programs, perhaps after reviewing the initial round of individual awards in a particular centers program, rather than on revisiting each center every three years. That would reduce the proposal review workload considerably. For example, 25 of the 31 research awards approved by the NSB in FY 1992 were renewals of individual centers in three programs (Engineering Research Centers [ERCs], Science and Technology Centers [STCs], and Materials Research Laboratories [MRLs]).

The NSB also could raise the delegation thresholds to concentrate on the very largest projects. Just adjusting the thresholds for inflation, since they were originally calculated in 1983 (but not adopted until 1986), would raise them to about $2 million a year or $8 million over five years.

DESIGNING THE REVIEW AND SOLICITATION PROCESS

NSF has had special procedural requirements for planning and conducting reviews of "large special projects" in its *Proposal and Award Manual* since 1987 (NSF, 1992a:I-7). These requirements, which must be approved by the assistant director and DARB, apply to competitions involving one or a few awards of $1 million a year or more that result from special program announcements or program solicitations.

NSB also must review and approve a summary document describing a proposed solicitation announcement, which states the goals of the project, prescribes the format for proposals, and describes the procedures and criteria that will be used to evaluate the proposal. The development and approval of solicitations are discussed below.

Proposal Review Planning Requirements

According to NSF review planning requirements, a plan and schedule for all significant review events must be approved by the cognizant assistant director and DARB.[2] The events to be planned

---

[2] The review planning requirements for major projects do not apply to unsolicited proposals and noncompeting renewal proposals. That category includes some long-standing national facilities such as the National Optical Astronomy Observatories; National Radio Astronomy Observatory; National Center for Atmospheric Research; Cornell Electron Storage Ring; nuclear

and scheduled include peer review deadlines, selection of panel members, panel meetings, site visits, DARB action, and NSB approvals. The plan and schedule are to be updated as necessary, and updates are to be approved according to the same procedures used for the initial plan.

Review formats are supposed to be structured according to the announced selection criteria for that specific competition and approved in advance by the cognizant assistant director. This is meant to ensure that reviewer comments address the criteria, that all criteria are covered, and that only the announced criteria are used.

Site visitors should have similar instructions, and the site visit reports should be structured so that criteria commented on in one report are also addressed in the reports on visits to other sites. If new items are added as a result of later site visits, appropriate addenda should be attached to earlier reports.

The summary review by program staff is supposed to be structured similarly to the review formats, (i.e., organized around the announced selection criteria). At every stage, the documentation of declinations should be comparable in level of detail to awards and proposals still in competition. Recommendations for awards are not supposed to be made until the review of all proposals has been completed. Finally, the review and selection process should be described in sufficient detail to show that NSF requirements were followed, including notes on telephone or other electronic communications with reviewers and proposers.

The timing and content of these procedural requirements indicate that they were responses to criticisms by the General Accounting Office (GAO) of the competition for the Earthquake Engineering Research Center (EERC). GAO found problems and inconsistencies in the way the reviews of the two leading proposals were handled,

---

physics facilities at Michigan State and Indiana; and the Ocean Drilling Program. It also includes more recent initiatives such as the Laser Interferometer Gravitational Wave Observatory and Incorporated Research Institutions for Seismology.

including reviewer comments that were not linked to the stated criteria and a preliminary decision to fund one proposal before the review of the other proposal was complete (GAO, 1987:Ch.3).

In practice, the directorates have fulfilled the 1987 requirement for formal planning of the review of large special projects in various ways. In any case, the requirement is procedural rather than substantive, focusing on the timetable for reviews and site visits rather than on consideration and justification of criteria, reviewer qualifications, or types of review mechanisms to be used. Any higher-level review and justification of the review process to be used for a major award takes place in the context of preparing the solicitation document (e.g., Request for Proposal, program announcement, project solicitation), which should be approved by NSB.

NSF recently adopted new internal guidelines for initiating new projects and programs—the "Design, Review, and Management Protocol" (NSF, 1993a). The new protocol calls for

- a demonstrated need and explicit goals toward which progress can be measured;
- a set of clear policies to guide the review process;
- adequate budget and staff to manage the proposal review process;
- management plan for funds and personnel that specifies responsibilities, with input from all participating units;
- closer coordination among administrative units involved in the grant-making process—the offices of grants and contracts, financial management, and information systems;
- criteria for measuring progress and an assessment plan for monitoring the impact of the activity;
- mechanisms for full and timely communication with the constituent research community and NSF staff; and
- the ability to be modified or improved in light of new information on progress, or discontinued if and when the goals are reached.

The initiating unit is directed to prepare a "management package" for approval at least at the assistant director level that includes, among other items, a review management plan outlining the process, criteria, deadlines, and administrative responsibilities to be employed in the competition. If the activity is subject to NSB review and approval, the management package is now expected to be an integral part of the background material supporting the decision. Also, program announcements, solicitations, "Dear Colleague" letters, and other external communications to the research communities that go through NSF's internal clearance process must be supported by an approved management package.

NSB Approval of Solicitation Announcements

In the 1970s, NSB began to require that it approve all formal announcements inviting proposals in which it was expected to decide on the eventual award (NSF, 1977). Among the other case studies, project announcements soliciting competitive proposals were approved by the NSB for the National Nanofabrication Users Facility (1977), ERCs (1984), Supercomputer Centers (1984), EERC (1985), and STCs (1987). In a related action, the NSB asked to review the site selection process and criteria for the two Laser Interferometer Gravitational Wave Observatory (LIGO) sites in October 1990.[3]

The project solicitations for the National High Magnetic Field Laboratory (NHMFL) and the recent recompetition of the National Nanofabrication Users Facility were not, however, formally reviewed by NSB.

Solicitation documents, although generated within the various directorates, have a fairly common format. The first "program announcement" for the ERC program, for example, began with a

---

[3] This review was required as part of NSB's approval of the award in May 1990 to a California Institute of Technology-Massachusetts Institute for Technology consortium for LIGO construction and operation.

description of the goal of the program as a whole, the defining characteristics of ERCs, and expected features of the centers (NSF, 1984a). These were based on a report of the National Academy of Engineering on guidelines for ERCs (NAE, 1983). The ERC program announcement also included information on who could submit proposals, deadlines, and expected award size and duration.

The announcement cited the four basic criteria used to evaluate all proposals (i.e., research performance competence; intrinsic merit of the research; utility or relevance of the research; and effect of the research on the infrastructure of science and engineering). It went on to say that "within these general criteria," consideration would be given to certain features in evaluating how well the proposed center might meet the objectives of the ERC program. These included such items as the importance of the research area addressed by the center; impact of the center on engineering education; industrial participation; cross-disciplinary nature of the center; and management plan.

ERC program announcements were modified each year. By 1989 the program announcement prescribed a more elaborate proposal format and a revised list of review criteria that subsumed the four basic criteria and included additional ones relevant to a university-based research center (NSF, 1988b:5): (1) research merit and potential impact on U.S. competitiveness; (2) strength and impact of educational programs; (3) industrial/other user participation and knowledge/technology transfer; (4) leadership and performance competence; (5) institutional environment and support; and (6) effect on the infrastructure of engineering.[4]

---

[4] A program solicitation was drafted for two ERCs to be funded in 1993, in the areas of advanced materials engineering and advanced manufacturing systems, but it was not issued for lack of funds (NSF, 1992c). It would have revised the criteria again by eliminating the one on institutional environment and support and adding one on "need of a center to accomplish the research program." This announcement would have been the first to indicate the priority of the criteria: "Criteria used to reach these judgments are listed below in their order of priority. Regarding decisions, the quality of the research is assessed first; if that is sufficiently high, then the quality of the other components of the ERC listed

The proposal solicitation for the NHMFL, issued by the NSF Division of Materials Research in late 1989, described a two-stage review process. In the first stage, proposals would be evaluated by mail and/or panel reviews. In the second stage, the institutions identified as having the most meritorious proposals in the first stage would be visited by a team of experts. "On the basis of the recommendations of this site visit team, the most highly meritorious proposal will be selected . . . and transmitted to the National Science Board for its review and approval" (NSF, 1989:3). The criteria included the four general criteria (with each explained in terms of the NHMFL) and two additional criteria[5]:

- the likely effectiveness of proposed management plans, and
- the level and nature of institutional and other sector commitments.

The program solicitation for STCs was based in large part on a report of a National Academy of Sciences (NAS, 1987) panel. It outlined a more elaborate two-tier review process:

The first stage of review will be conducted by NSF's research directorates, and will focus particularly on the scientific aspects of the proposals . . . . Only those proposals deemed most competitive during this scientific review stage will be reviewed further in the STC competition.

The second stage of the review process will involve a comprehensive review by a multi-disciplinary, NSF-wide

---

below enter into the decision."

[5] It has already been noted in Chapter 3 that the 50 percent nonfederal matching requirement specified in the project development plan turned into a requirement for "substantial" cost sharing in the solicitation announcement.

panel specially convened for the STC program . . . . In addition to considering the criteria discussed below and elsewhere in this announcement, the NSF-wide panel will examine the balance of awards among scientific fields and their combined ability to meet the goals of the NSF STC Program, including enhancing the Nation's economic competitiveness.

The criteria for selecting STCs were similar to those for ERCs and other center programs (e.g., intrinsic merit, competence, utility or relevance, and infrastructure impact, plus appropriateness of the center approach, management plan, educational effects, and private sector linkages and knowledge transfer arrangements), and as with the ERCs, no indication was given of their ranking.

Other project solicitations examined by the panel had a similar format: a page or two describing the project goals, and sections on who may apply, responsibilities of the principal investigator, deadlines, proposal format and content, evaluation criteria, and the award size and instrument. Most referred to the four general NSF criteria and also listed additional ones. Only a few indicated the priority of the criteria.[6] Most described the review process in a short general paragraph that gave maximum discretion to NSF: proposals will be evaluated by a combination of peer review, panel review, and site visits; proposers may be asked for additional information; and proposals may be rejected any time after the initial peer review. The solicitations for recent multidisciplinary ERC and STC programs have announced more elaborate multistage review processes, but the

---

[6] In addition to the 1989 solicitation for ERCs described above, these included the project solicitation for management and operation of the Sondrestrom incoherent-scatter radar facility in Greenland. It listed three primary criteria "equal in importance": (1) capabilities of principal investigator and staff, (2) technical/logistic support, and (3) scientific research program, and two secondary criteria: (4) educational potential and (5) management plans.

priority of criteria at each stage has not been articulated, except in the unissued 1992 version of the ERCs solicitation announcement (NSF, 1992c).

Findings and Recommendations on Proposal Review Planning

Recommendation 8: Planning the Review Process and Criteria

NSF and NSB should further strengthen their effort to implement a review process for each major award that (a) imposes a reasonable schedule, (b) identifies the appropriate selection criteria and their relative priority, (c) uses the two-phase review process, (d) selects appropriate reviewers to address each criterion at each stage, and (e) is assisted by a central office for review of major projects that ensures quality and consistency based on extensive experience with such complex project reviews.

a. *Reasonable Schedule*: The deadlines for the proposal and review process should leave adequate time for proposers to prepare proposals and for reviewers to do their jobs, and should be specified in the solicitation.

b. *Criteria and Their Priority*: NSF should make explicit the criteria and their priority in advance of the solicitation. The primary technical criteria that will be used in the first phase and the other criteria to be taken into account in the second phase should be identified at this time. Explicitness would also make it easier to identify appropriate reviewers to address the criteria at each stage.

c. *Two-Phase Review Process*: The details of the review process to be used for this award should be determined and specified in the solicitation document: how the first-phase technical review and second-phase overall review will be structured and scheduled.

d. *Appropriate Reviewers*: The reviewers with relevant expertise should be identified for each phase of the review process, and plans should be made to carry over some of the first-phase technical reviewers to the second-phase panel to participate in identifying the best proposal overall (and to ensure that technical quality remains the primary criterion).

e. *Learning from Experience*: NSF should have a central review office for major awards—an "institutional memory" that ensures consistency and learning from experience in designing and managing the review processes for major awards.

The panel found that the proposal review and selection process has been flexible and varied. This might be appropriate for individual grants, but it can be too inconsistent and easily misunderstood for major awards. The review phase is sometimes confusing and not always well understood by proposers or other observers. The design and implementation of the proposal review process for major projects should be done with special care, should follow directly from the stronger and more detailed project development planning process recommended in Chapter 2, and should be communicated fully to the research community and public.

NSF has procedures for designing an appropriate review process for new programs and projects, including major project awards, and these have recently been strengthened. The new protocol adds requirements for ensuring adequate resources to conduct the review process and for considering explicitly the conditions for discontinuing the project (see Chapter 5).

The goal of the review planning exercise is to ensure that the criteria, procedures, types of reviewers, use of site visits and review panels, involvement of the directorate advisory committee, and other aspects of the review are well thought out and appropriate. The panel strongly endorses this approach and urges NSF to implement the new policies fully. They should be closely linked to the processes for obtaining NSB approval of the project development plan and of the program announcement or solicitation that follows the plan.

A carefully prepared solicitation based on a thorough plan for the review could go far to head off subsequent misunderstandings and conflicts when an award is made. As described in Chapter 3, the project development plan for the NHMFL explicitly called for a 50 percent match of the then-estimated project cost of $105 million, not counting capital costs of construction or renovation of a building. The solicitation announcement only said that the facility was intended to be "heavily cost-shared" between NSF and other federal, state, or private sources. NSF gave the award to the proposal that had the 50 percent match (including $58 million from the state government). Recently, NSF decided to recompete the National Nanofabrication Users Facility that a university had operated with NSF support since it won the original competition in 1978. Although a national user conference had called for additional facilities, NSF decided to recompete the existing facility in 1992; thus, the solicitation called for proposals to manage *a* facility. NSF received several strong proposals and recommended two to the NSB for approval (including the incumbent). At this time, NSB heard from other institutions that they had decided not to apply because they thought the odds of succeeding against an incumbent were too low, but they probably would have applied if the solicitation had said that more than one facility would be awarded. As a result, NSB decided to have another competition in which the solicitation said that NSF intended to make more than one award.

There is no guarantee that discussion of the solicitation announcements by NSB would have avoided these situations, Nevertheless the requirement for NSB review and approval at the solicitation stage would give that body, with its broad outside perspective, an opportunity to identify potential issues and misunderstandings.

DOCUMENTING AWARD DECISIONS

According to NSF's current procedures, at the time of the decision to recommend funding or to decline the proposal, the proposal folder should contain

- the review plan required for large special projects;
- the request-for-review letter that refers to the selection criteria and their relative importance;
- the ad hoc mail reviews;
- advisory committee/review panel reviews and summary;
- site-visit reports;
- responses of the principal investigator to review comments or program questions; and
- correspondence, memoranda, or diary notes relating to the recommendation.

The program officer then completes the "Review Record" (NSF Form 7), which accompanies every recommendation for a new award, renewal, or declination or any other action subject to peer review. The form was developed in response to the curriculum development controversies in 1975. The purpose of the form was to document who the peer reviewers were and how they rated the proposal, and to ensure that the program officer responded to the concerns of each reviewer, especially those not consistent with the program officer's award decision.[7]

The first part of Form 7 lists the name, sex, department or field, and institutional affiliation of everyone asked to review the proposal, and indicates the summary rating (letter or number) given

---

[7] In 1975, congressional critics of NSF education activities discovered that an NSF program officer had selectively quoted the favorable sections of several critical peer reviews in justifying an award for a curriculum development project.

by the reviewer. If external review was required but there were fewer than three mail reviewers or panelists, the program officer must justify making a recommendation on the basis of one or two reviews.

If a review panel assigned an overall rating, it is recorded on the form.[8]

The program officer gives a summary rating of the reviewers' comments on previous NSF-supported work (and may indicate his or her rating if different from the reviewers').

The second part of Form 7 is the "review analysis," in which the program officer justifies the recommendation. If the recommendation is negative, "excellent" review ratings must be explained. If the recommendation is favorable, the program director must explain any "fair" or "poor" ratings or significant negative comments by reviewers.

After the program officer signs the completed Form 7 and prepares an abstract, a summary budget, and an administrative processing form, the proposal jacket is forwarded for review and approval via the section head (if there is one) to the division director. The division director is usually the final sign-off authority for the traditional small grant. The section head and division director are supposed to determine that the number and quality of external peer reviews were adequate, that significant peer review comments contrary to the recommendation have been dealt with adequately, that the rationale for the recommendation is reasonable, and that proper administrative procedures have been followed.

In the case of major awards, these determinations are made by ARG and DARB (as explained earlier). After reviewing the decision package going to CPP and EHR and to NSB—which contains the proposal, peer reviews, site visit and panel reports, and program

---

[8] It should be noted that Recommendation 6 calls for two panels for every major award. The first panel would give a summary rating and ranking of each proposals for technical merit (Phase 1) and, for those that pass Phase 1, the second panel would give a summary rating and ranking for overall merit (Phase 2).

director's recommendation—DARB reviews the memorandum prepared by the directorate for the director, recommending that NSB approve the award. The director's memorandum is normally a few pages that "summarize strategic information and issues on the proposed action, including objectives, alternatives, potential policy implications, precedents involved, and any other factors that could be considered nonroutine" (NSF, 1992a:III-2). It should have a short statement responding to any "major concerns" raised by reviewers, a summary of budget totals, the percentage of the program or division budget involved in the award, and the out-year budgetary implications.

The director's memorandum is the only public document laying out the basis and rationale for a major award. Currently, the review process for major awards is modeled closely on the peer review process for individual research projects, in which confidentiality is deemed necessary. Accordingly, the peer reviews and the site visit and panel reports are confidential. In most cases, the director's memorandum also discusses only the winning proposal because unsuccessful proposals are deemed confidential. Except in the case of the NHMFL award, in which the top two proposals were explicitly compared, the basis for the award is never comparative, which limits NSB and public understanding of the decision.

Findings and Recommendations on Award Documentation

<u>Recommendation 9: More and Better Public Documentation of Award Decisions</u>

The review and award process should be fully documented and the results made more accessible than is standard or necessary for traditional individual investigator proposals. This process includes such documentation as site visit and panel reports, and the staff-prepared director's memorandum to the NSB summarizing the review results and recommending the awards. In particular, as

recommended above, any decision to pass over the proposal rated highest technically (Phase 1) or to recommend a proposal other than the one selected in Phase 2 of the merit review process must be fully explained, and relevant documents should be publicly available.

Because of their size and importance in a field, major project awards are more significant and more public than small grant proposals submitted by individuals and small groups. They also usually involve multiple criteria and complicated choices whose basis is harder to understand than purely technical merit in making individual research projects grants.

As noted in Chapter 3, the panel found that the criteria and procedures are not always well understood by participants, including proposers and peer reviewers, or by interested observers. They are sometimes not sure of the relative importance of the various criteria or how to fulfill them, and they do not always understand how the criteria are eventually integrated by staff in making the final decision. Certain stages of the review process are not well documented, at least publicly. This is an additional source of misunderstanding of the basis for final decisions.

The panel concluded that it would be very beneficial for NSF staff to be more explicit and open in explaining and documenting recommendations to NSB and for NSB to be more explicit in documenting the basis for its approvals.

The most appropriate vehicle to explicate major award decisionmaking is the director's memorandum recommending a proposal for funding to NSB. This memorandum should be written with public dissemination and understanding in mind.

After making its decision, NSB also should issue a statement of its reasoning for approving the award, because it may amend or differ from the rationale contained in the director's memorandum. These documents would make the decisionmaking more understandable in some cases if there were an analysis of the strengths and weaknesses of the approved proposal, compared to those of the runner up or other leading proposals. This is especially necessary if a proposal other than the one ranked highest in the first-phase technical review is recommended for funding. The comparison would follow more easily if the earlier recommendation were adopted that there be an advocate for the top technical proposal in such cases in the final decision-making by NSB.

We are not recommending that the names of peer reviewers, or other information that might dampen reviewer candidness, be made public. The directors's memorandum is supposed to summarize reviewer comments; we believe that the documentation of the NSB award decision and its reasoning can be done well without disclosing the identity of individual peer reviewers or review panel members.

To ensure wide access to the documentation, NSF could use Internet to make the documentation for major award decisions readily available on-line through its Science and Technology Information Systems, as it already does abstracts of winning award proposals. NSF also could make major award policies and procedures more available electronically. NSF recently made the *Grant Policy Manual* available electronically and updates it periodically. The *Grant Policy Manual* is intended primarily to provide grant administration requirements to those who have received small grant awards (it does not, for example, mention the NSB role in reviewing large awards). NSF's (1993c) internal *Proposal and Award Manual* contains the full set of merit review and award policies and procedures for major awards, and the relevant sections should be made available electronically and updated regularly.

5 Recompetition of Awards

PROJECT CONTINUATION AT NSF

Projects receiving major awards are usually long-term activities supported by multiyear continuing grants or cooperative agreements. These activities require continuing oversight by the National Science Foundation (NSF) and its programmatic advisory committees. Periodic decisions must also be made on whether or not to renew support when the current grant or cooperative agreement expires, which occurs at least every five years. The ongoing evaluation process typically involves annual reports from the project, visiting committees, and site visit evaluations by NSF staff and advisors. This may lead to conditions being placed in the grant or contract renewal that require programmatic or managerial changes. This ongoing evaluation process occasionally leads to terminations or opening of a renewal award to competition.

Several internal and external postaudit processes are in place to ensure that NSF policies and procedures for handling proposals and making awards, are followed. These processes include in their scope major awards, although they are not treated separately or specially. For example, major awards are reviewed as part of the committee-of-visitors evaluation of each NSF program that takes place on a three-year cycle. Also, the director makes an annual report to the National Science Board (NSB) on the performance of the peer review process. The Office of the Inspector General audits a sample of awards each year to determine the level of compliance with official policies and procedures. From time to time, the General Accounting Office audits a particular decision (e.g., the Earthquake Engineering Research

Center [GAO, 1987] or the home basing of oceanographic research vessels [GAO, 1982]).

Many of the major projects, involving most of the major award funding, come up for noncompetitive renewals at the end of each grant or contract period (usually three to five years) and are expected to be continued if their performance has been satisfactory. These include both the large national facilities managed by consortia of research institutions and the national user facilities managed by individual universities (e.g., the National Optical Astronomy Observatories at Kitt Peak, Arizona, and near Cierra Tololo, Chile; National Radio Astronomy Observatory at Green Bank, West Virginia; National Astronomy and Ionosphere Center's radio/radar telescope at Arecibo, Puerto Rico; National Center for Atmospheric Research; the Incorporated Research Institutions for Seismology's (IRIS) global seismometer network; oceanographic centers, vessels, and other facilities; the Cornell Electron Storage Ring and various nuclear physics facilities; and Supercomputer Centers. In addition, the following new facilities under construction will be managed on an open-ended basis (no sunset period specified): the GEMINI telescopes; the new Green Bank radio telescope; the National High Magnetic Field Laboratory (NHMFL); and the Laser Interferometer Gravitational Wave Observatory (LIGO).

Decisions must be made periodically on what to do when the current grant or cooperative agreement for a major project expires. At the end of the specified grant period, NSF should make several decisions:

1. *Is this still a worthwhile activity or has it become obsolete in the face of scientific and technological advances?*

In the 1970s, for example, NSF withdrew support from a number of on-campus nuclear reactors and accelerators, converting some of the projects to research groups using national nuclear and high-energy physics facilities. In the 1980s, NSF superseded its support of campus computer centers with regional supercomputer centers that can be accessed remotely by all researchers. As supercomputers become cheap enough for every campus to afford, supercomputer centers may in turn become obsolete.

To determine whether an activity is still worthwhile, NSF must evaluate program relevance and priority. This aspect of major project decisionmaking is not well developed (it is a continuation of the planning process described in Chapter 2, which our recommendations in that chapter would strengthen and make more explicit), compared to NSF's more elaborate procedures for deciding whether or not a particular facility or center is worth continuing or should be competed again.

2. *If the activity is still deemed worthwhile, is the current grantee doing a good job that merits continuation or should the award be opened to competition?*

This question is addressed primarily by the usual proposal review policies and procedures of the NSF and NSB. This process works reasonably well, although we make recommendations in Chapters 3 and 4 to strengthen it. Since most projects can show evidence of being productive and still worthwhile, they can justify renewal. The NSF and NSB should have procedures in place to ensure, therefore, that a facility, center, or other major project is being operated by the best possible grantee.

Recently, NSF has begun to emphasize such procedures in the conditions it sets for renewing or recompeting awards. The Engineering Research Center (ERC) awards, for example, have been made subject to elaborate renewal procedures every several years and have an absolute sunset provision of 11 years, after which a center is on its own or must recompete on an equal basis with new proposers. Science and Technology Center's (STCs) are subject to a similar procedure.

Sunset provisions have been built into other awards recently. The last time the Ocean Drilling Program (ODP) was renewed, for example, the NSF and NSB reviewed and approved a 10-year plan for the period beginning in FY 1994; approved a five-year contract using the current drill ship; called for review and renegotiation of the contract after 1998 to accommodate a new or additional drill ship if needed; and stated NSF's "intent to terminate the ODP by the end of FY 2003" (NSB, 1992).

Unique national user facilities operated at and by a particular institution constitute a class of major awards that has been especially difficult to compete and recompete. They may provide a great benefit to the host institution and community (or loss if another place wins the recompetition), involve sunk costs in the existing facility that may be lost if it is terminated, and pose significant transition costs for the national user community while the new facility gets off the ground. Nevertheless, NSF has begun to recompete some of the university-based national facilities previously considered open-ended, including the National Nanofabrication Users Facility and the NHMFL. It is planning to recompete the Arecibo radar/radio telescope facility operated by Cornell.

In 1986 an NSF staff task force studied policies and procedures for terminating programs and major projects. They documented examples of successful terminations but also identified factors that tend to impede termination (e.g., the inherent interest of the peer review system, advisory committees, and program officers in identifying expansion areas and new opportunities; the development of constituencies for established activities and consequent loss of interest in other options or modes of program activity). The task force recommended a comprehensive but flexible system of sunset review of all major activities and the articulation of termination plans and contingencies as part of the initial planning (cited in NSB, 1988a:32).

In 1988 an NSB committee on centers and individual investigator awards addressed program termination as a possible way to ensure funding new people and activities even with steady or declining budgets. The committee's guidelines for orderly termination of projects no longer needed were based on competitive proposal review to ensure that the most meritorious projects were supported (NSB, 1988a).

In 1991 the NSB considered adopting a time limit on all continuing activities (NSB, 1991b:2). Although the sense of the meeting was that major projects should be reconsidered every 10 years, NSB stopped short of setting a specific number: "Automatically recompeting a major center or research facility every five years does not appear reasonable; however, renewal proposals undergo a rigorous peer review, typically including a site visit. Continuity of support is needed for at least 10 years, provided the facility or center is performing in a satisfactory manner." NSB also decided that "when recommendations for major projects are presented to the Director's Action Review Board for waiver or explicit consideration, a statement regarding plans for the end of the grant period will be included. Further, changes in the renewal plan can be made as appropriate; however, any deviation from the renewal plan will be brought to the attention of the appropriate NSB committee." Currently there is no strict sunset requirement, although there is now a sense that major projects should be reconsidered at preset intervals and at least every 10 years.

Finally, as described in Chapter 3, the director recently approved a new program initiation protocol that requires among other things that the expected duration of a new program be specified (NSF, 1993a). The protocol also calls for a monitoring and evaluation plan to determine when an activity should be discontinued because it is no longer effective or needed.

Findings and Recommendations

Recommendation 10: More Recompetitions

The initial planning of every major award should specify the conditions for renewing, recompeting, or terminating the project. As a general rule, each project (or perhaps, in the case of large

national facilities, its management) should be recompeted openly within a time period appropriate to the nature of the activity. Such periodic recompetitions should be preceded by an assessment of whether such an activity, however successful, is still needed or whether the funds would be better used in research areas of higher priority or for other mechanisms (e.g., grants to individual investigators instead of a research center, or a program of university instrumentation awards in place of a central national facility).

The panel believes that periodic competition results in the highest-quality proposals and that grantees will perform better knowing that eventually they will have to defend their stewardship of a major center or facility. This benefit justifies the extra costs of a periodic competition. This recognition should be built into the solicitation of major awards whenever possible. This might extend to the management of large national research facilities, although it may be very difficult to find a competitor when the project is managed by a national consortium of universities engaged in the area of research, such as the University Consortium for Atmospheric Research (UCAR) and the Associated Universities for Research in Astronomy (AURA), or a facility is located at a single university and cannot be moved, such as the Cornell Electron Storage Ring.

In awards for unique national facilities that can be competed, the grantee should be required to agree from the beginning to cooperate fully in a transition to a new operator if the original operator should lose a recompetition. For its part, NSF should develop and NSB should adopt uniform guidelines for orderly transitions so that facility operators know what to expect should they lose a recompetition some day.

Since an existing facility tends to have a natural advantage over a proposal on paper—the existing facility is operational, has staff and a track record, and sunk costs have been paid already—NSF should make every effort to create a level playing field. In establishing or relocating major facilities, however, it is very difficult accurately to estimate transition costs and the loss of momentum in ongoing activities. This is especially true in rapidly moving fields, which are often the ones most likely to need new investments and institutions. The problem of underestimating these costs must be balanced against tilting the competition toward the incumbent by overestimating transition costs. This means that in recompetitions, the impacts on users and transition costs should be fully considered and included in the budget plan, especially when it will take time for a new facility to become operational. An agreement also should be worked out with the Office of Management and Budget (OMB) and the appropriations committees as to how to handle budgeted transition costs if the incumbent grantee wins the recompetition.

The problem is that if the transition costs go to the successful grantee, the user community and NSF benefit from renewing the incumbent, which may bias the competition. On the other hand, if realistic transition costs are not taken into account, the bias will be against the incumbent. Given that the existing center or facility has a natural advantage, the purpose of a recompetition should be to seek better proposals that would justify the extra costs of relocation.

The adequacy of support should also be carefully assessed as part of the periodic review of the continuing programmatic need for a major project, which we recommend should take place before considering whether or not to renew or recompete any award. There is a natural tendency in launching new centers programs, for example, to respond to budget pressures by funding each center at lower levels than expected in order to get as many centers started as possible. This happened with the ERC and STC programs. In programs with multiple centers or regional facilities, therefore, NSF should periodically assess the adequacy of funding of each center. Despite pressures to keep every center going, NSF should make adequate funding of each center it sponsors a priority even if it has to cut the overall number of centers.

Maintenance and upgrading are also often the first to go when funding is tight. In the case of large-scale physical facilities, therefore, NSF should take care to ensure adequate maintenance and upgrading. Even with significant real budget growth in the late 1980s, existing facilities did not always receive adequate funding for maintenance and incremental upgrades. The usability of the Very Large Array, for example, had seriously declined. In response to this and similar problems with other astronomy facilities, the Astronomy and Astrophysics Survey Committee of the National Research Council made "restoring the infrastructure" of existing equipment its highest priority (NRC, 1991).

6 *Looking to the Future*

This report concludes that merit review—the peer review-based system of the National Science Foundation (NSF) for making awards—has generally worked well in decisionmaking on major projects over the years. It also concludes that the merit review system needs to be adjusted in certain ways to meet contemporary conditions.

A major theme of the report has been that NSF should plan very carefully what to do, and how to do it, before worrying about specific proposal selection—because the key programmatic decision is whether or not to do something, not who eventually will get the award to carry it out. Such careful analysis of program needs and opportunities should first involve outside advice and staff judgment and then careful consideration by the National Science Board (NSB); it should precede and help shape the solicitation and review processes.

Greater emphasis on front-end thinking will be increasingly important in the future because most fields of science are undergoing revolutionary change. There are exciting new discoveries, new more powerful instrumentation and facilities, and computational power is expanding dramatically. Advances in communications technologies are fostering the use of teams of researchers—even those widely separated geographically. As a result, all of science—both large and small, interconnected and individualistic—is becoming more dependent on expensive instruments and facilities. Rising costs also mean that support for some awards for unique facilities may have to come from other nations. Finally, actions taken by a federal government agency must increasingly be made with an awareness of investments by state governments, for example, in supercomputing facilities.

The panel was not asked to address the issue of the appropriate balance between "big" and "little" science. This has been a contentious issue because of the fear that large projects may reduce the support available for the type of small research projects conceived and conducted by individual scientists and engineers that have led to many important scientific advances and breakthroughs. The big-little balance has become very serious due to a combination of tight NSF budgets and the relative appeal of large programs to Congress and the public over the myriad of small projects. Large projects are often the best research investment that can be made, but before this conclusion is reached in any specific field, a careful analysis should be carried out of the potential impact that funding of such a project or projects is likely to have on the overall productivity of research supported by NSF in that field. There is no overall answer to the balance question, because it differs from field to field and will vary over time within each field in response to new discoveries and the availability of new instrumentation.

We believe that it is desirable to find new strategies for dealing with major awards that take into account the increasing interrelatedness of big and small modes of research. Such strategies probably would involve field-level reviews of all modes taken together. The recent National Research Council (1991) report, *The Decade of Discovery in Astronomy and Astrophysics*, supported by NSF, was such an effort; it looked comprehensively at the maintenance and upgrading of existing facilities, the need for new initiatives, and individual project support and training of new researchers within a specific area of science. In an earlier example, planning conducted jointly by NSF with the earth sciences research community resulted in the Continental Lithosphere Program. The program balanced big science facilities, such as the global seismic array and the continental scientific drilling program, with support for individual investigators to use those facilities and conduct complementary, small-scale research in the field. This approach also helps meet the need to use systematic field-by-field assessments for setting budget priorities (NAS, 1989) and measuring national performance in research (NAS, 1993:Ch.3).

In thinking about major projects, NSF and NSB should consider using regular and comprehensive peer reviews of research fields more widely to determine the appropriate interactions among modes of research and the most productive role of major projects within the fields. Major awards should then be justified on the basis of whether or not they contribute to the most productive mix of research and the overall health of the research field. In this way, the overall strategy for a field would be more consistent internally, more understandable to the affected research community, and also more intelligible to the public and its elected officials who ultimately bear the responsibility for investment in research.

At the same time, NSF and NSB should be ready to change plans flexibly in light of new developments. Planning assumptions in a fast-moving field of research become obsolete quickly as they are altered by the progress of research. Field reviews should be periodic and adjusted regularly as part of NSF-NSB's long-range and annual planning process. Each major project should always receive special scrutiny because, although it may provide unique opportunities to conduct certain kinds of research, it reduces NSF's flexibility to fund research opportunities that did not even exist when the project was approved.

A Biographical Sketches of Panel Members

ROBERT H. RUTFORD, Ph.D., the panel's Chairman, is President of the University of Texas at Dallas and Professor of Geosciences. He is also Chairman of the Polar Research Board of the National Research Council. Dr. Rutford holds the National Science Foundation's Distinguished Service Medal and the Antarctic Service Medal, and he was previously Director of its Division of Polar Programs. He is the U. S. Delegate to the Scientific Committee on Antarctic Research, a Fellow of the Geological Society of America, and a member of the Board of Trustees of Baylor Dental College, as well as a member of a variety of other community and professional boards and committees.

CLARENCE R. ALLEN, Ph.D., is Professor of Geology and Geophysics, Emeritus, California Institute of Technology. Dr. Allen has been President of the Seismological Society of America and the Geological Society of America. He is a member of the American Association for the Advancement of Science, the American Geophysical Union, the Geological Society of America, the National Academy of Sciences, and the National Academy of Engineering. Dr. Allen received the first G.K. Gilbert Award in Seismic Geology.

ALBERT A. BARBER, Ph.D., Special Assistant to the Chancellor, University of California, Los Angeles (UCLA), acts as the university liaison with federal agencies and higher education associations. He was formerly Vice Chancellor—Research Programs and Chairman and Professor of Zoology at UCLA. Dr. Barber chairs the Board of Directors of the National Association for Biomedical Research and is a member of the American Association for the Advancement of Science, the American Physiological Society, the American Society of Biochemistry and Molecular Biology, Phi Beta Kappa, and Sigma Xi.

HARVEY BROOKS, Ph.D., is Gordon McKay Professor of Applied Physics, Emeritus, in the Division of Applied Sciences, and Benjamin Peirce Professor of Technology and Public Policy, Emeritus, at the John F. Kennedy School of Government, Harvard University. He is a former member of the President's Science Advisory Committee and the National Science Board and was President of the American Academy of Arts and Sciences. Recently he was a member of the Advisory Council of the Carnegie Commission on Science, Technology, and Government and served on four of its task forces. Dr. Brooks is a member of the National Academy of Sciences, the National Academy of Engineering, the Institute of Medicine, and the American Philosophical Society.

CHRISTOPHER COBURN is Director of Public Technology Programs, Battelle Memorial Institute. He is also Staff Director of the Carnegie Commission on Science, Technology, and Government's Task Force on Science and Technology and the States. Formerly, he served as Executive Director of Ohio's Thomas Edison Program and Science and Technology Advisor to the Governor of Ohio. He founded and chaired the Science and Technology Council of the States.

SUSAN E. COZZENS, Ph.D., is Associate Professor in the Department of Science and Technology Studies at Rensselaer Polytechnic Institute and Director of Graduate Studies for the department. She was formerly a policy analyst in the Division of Policy Research and Analysis of the National Science Foundation. While at NSF she also served as Associate Executive Secretary of the Director's Advisory Committee on Merit Review and as a consultant in the review and reorganization of its program evaluation activities.

Dr. Cozzens is outgoing editor of *Science, Technology and Human Values*, the journal of the Society for Social Studies of Science.

FRANK D. DRAKE, Ph.D., Professor of Astronomy and Astrophysics, University of California, Santa Cruz, previously served as the university's Acting Associate Vice Chancellor, University Advancement, and Dean, Natural Sciences Division. Dr. Drake is a former Director of the National Astronomy & Ionosphere Center, which includes the Arecibo observatory. He has chaired the U.S. National Committee for the International Astronomical Union and the Division of Planetary Sciences of the American Astronomical Society, and has been the Chairman of the Board on Physics and Astronomy of the National Research Council. He is a member of the National Academy of Sciences and the American Academy of Arts and Sciences.

DONALD S. FREDRICKSON, M.D., is President of D.S. Fredrickson, Inc., an international consulting firm, and a part-time Scholar of the National Library of Medicine, engaged in historical research on the support of biomedical research. Dr. Fredrickson was formerly Director of the National Institutes of Health and President of the Institute of Medicine. More recently, he served on the White House Science Council and as President and Chief Executive Officer of the Howard Hughes Medical Institute. Dr. Fredrickson is a member of the Institute of Medicine, the National Academy of Sciences, the American Philosophical Society, and the American Academy of Arts and Sciences.

FREDRICK S. HUMPHRIES, Ph.D., has been President of Florida A&M University since 1985. He was previously President of Tennessee State University. Dr. Humphries currently serves on the Commission of the Future of the South, the Science and Technology Advisory Committee of NAFEO, the White House Science and Technology Advisory Committee, and the State Board of Education Advisory Committee on the Education of Blacks in Florida, which he chairs. He holds the Meritorious and Distinguished Achievement in Education Award, Nashville Chapter, and the Distinguished Service to the Advancement of Education for Black Americans Award, among others.

ANITA K. JONES, Ph.D. (*NOTE*: Dr. Jones resigned from the panel on May 31, 1993, to become Director of Defense Engineering, Department of Defense, and did not participate in drafting the report after that date.) Until going to the Department of Defense, Dr. Jones was Professor and Chair of the Department of Computer Science, University of Virginia, and Editor-in-Chief of *Transactions on Computer Systems*, a quarterly journal. Previously she founded and served as Vice-President of Tartan Laboratories, Inc. Dr. Jones has been a trustee of the MITRE Corporation, member of the Air Force Science Advisory Board, the Lincoln Laboratory Advisory Board, and the Defense Science Board. She has participated as the chair or member of numerous program committees for computer science conferences and has served as an officer in several professional organizations.

LARRY K. MONTEITH, Ph.D., Chancellor of North Carolina State University, also served as Interim Chancellor, Dean of Engineering and Head of the Department of Electrical Engineering. Prior to his appointments at North Carolina State, Dr. Monteith was head of the Materials and Devices Laboratory at the Solid State Laboratory of the Research Triangle Institute and a member of the Technical Staff of AT&T Bell Laboratories.

DOUGLAS D. OSHEROFF, Ph.D., is Professor of Physics, Stanford University. He was previously the Head of the Solid State and Low Temperature Physics Research Department at AT&T Bell Laboratories. Dr. Osheroff holds the Walter J. Gores Award for Excellence in Teaching and the MacArthur Prize Fellow Award, among others. He is a member of several professional associations, including the American Academy of Arts and Sciences, the American

Physical Society, and the National Academy of Sciences, and also serves as Secretary of the International Union of Pure and Applied Physics Commission on Low Temperature Physics.

JUDITH A. RAMALEY, Ph.D., President and Professor of Biology at Portland State University, has held faculty and administrative positions at Indiana University, the University of Nebraska, the State University of New York at Albany, and the University of Kansas at Lawrence. She was Chair of the Academic Affairs Council of the National Association of State Universities and Land Grant Colleges, and Chair of the Commission of Women in Higher Education of the American Council on Education. She is a charter member of the Advisory Committee for the Biological Sciences of the National Science Foundation and also serves on a variety of professional boards, committees, and associations.

LYLE H. SCHWARTZ, Ph.D., was appointed to his current position as Director of the Materials Science and Engineering Laboratory of the National Institute of Standards and Technology in October 1984. Previously he served as Professor and then as Director of the University Materials Research Center at Northwestern University. Dr. Schwartz chaired the panel on international competition and cooperation of the Materials Science and Engineering Study of the National Research Council, and chairs the intergovernmental Committee on Materials Science and Engineering (COMAT). In 1990 he received the Presidential Rank Award of Meritorious Executive for outstanding government service.

B Major Awards Supported by NSF

This appendix includes (1) a typology of major awards by mechanism (center, facility, etc.); (2) an estimate of how the appropriation for the Research and Related Activities budget of the National Science Foundation (NSF) is distributed by mechanism; and (3) an overview of how the major awards are distributed among the NSF research directorates.

TYPOLOGY

Center Programs

NSF funds about 60 centers large enough to fall into the major award category. The rationale for supporting research centers at universities is to focus on complex scientific and engineering problems that need more expensive facilities and equipment, longer-term support, and larger-scale (usually interdisciplinary) attention than grants to individual investigators or small groups of researchers. It should be noted that the individual investigators associated with centers may and often apply for and receive support from standard NSF research grants and from other agencies for part of their work. Similarly, many of the facilities exist mostly if not completely to provide access to expensive instruments to individuals and small groups who could not otherwise afford to have them.

<u>Materials Research Laboratories</u>. The first centers program supported by NSF, the Materials Research Laboratories (MRLs) set

the pattern. The MRLs were originally established as "Interdisciplinary Laboratories" by the Advanced Research Projects Agency in the early 1960s to foster sustained interdisciplinary research on materials using costly, sophisticated equipment (Sproull, 1987). When the centers were transferred to NSF in 1972, the emphasis on interdisciplinary work was increased (Schwartz, 1987). As NSF (1973) put it at the time, "scientific excellence is viewed as a necessary but no longer sufficient condition to qualify for MRL core support." The majority of funding was expected to go to "coherent multi-investigator projects in major thrust areas requiring the expertise of two or more materials-related disciplines," and the MRL proposals also had to meet additional criteria, including effectiveness of local management, extent of support by university administration, level of interdepartmental cooperation, amount of education and training, and fit of proposed research areas within the overall program (NSF, 1973:3). Today there are 9 MRLs, including 8 of the original 12 (several others have entered and left the program since 1972).

Engineering Research Centers. In the 1980s, NSF launched two large center programs for engineering research and science and technology research. The 18 Engineering Research Centers (ERCs) are campus-based interdisciplinary research centers focused on problems related to national economic competitiveness. The program's goals, design, and proposal review process were based on advice from the National Academy of Engineering (NAE, 1983, 1984). As with the MRLs, there are important criteria in addition to the technical excellence of the research proposed. These additional criteria include the contribution of the center type of organization to sustained interdisciplinary research on relevant problems, the degree of cost sharing by state government and industry in order to promote relevance of the research to eventual industrial users, and the impact on education and training. The review process prior to award was elaborate. Each proposal was sent out by mail for review to at least six experts, followed by a panel meeting to identify proposals worthy of a site visit. After the site visits, the review panel met again to

APPENDIX B

choose the best proposals. The NSF staff then recommended awards to the top-ranked proposals as far as the funding went, and the National Science Board (NSB) reviewed and approved them. The first six ERCs were funded in FY 1985, five more in FY 1986, three in FY 1987, three in 1989, and four in FY 1990; three of these were terminated after their first five-year award.

Science and Technology Research Centers. The Science and Technology Center (STC) program grew out of a presidential initiative to foster basic research in areas of potential significance to national economic competitiveness. The first 11 STCs were funded in FY 1988, and 14 more were started in FY 1990. As with the ERCs, the proposal review process was elaborate, involving mail reviews, site visits, and panels to winnow down the numbers to a small group of finalists. As in the other center programs, factors other than the technical quality of the proposed research per se were important.

Other Major Research Centers. The MRL, ERC, and STC programs account for 52 of the 59 major centers subject to NSB approval. The other seven include the Earthquake Engineering Research Center (EERC) (1986), three biological research centers (1988), a plant science center (1988), the National Center for Geographic Information and Analysis (1989), and one of the 50 Industry/University Cooperative Research Centers (1992) (the other 49 are too small to require NSB review).

Centers represent a class of major awards that grew rapidly in the mid- to late 1980s. They constitute the majority of awards that the NSB must review and approve each year because they are supposed to undergo full merit reviews every three years in order to receive a new five-year award.

National User Facilities Run by Consortia

The oldest and largest facilities supported by NSF are managed by consortia of the institutions most involved in the relevant field of research. The costs of operating these facilities account for a large share of the funding for major awards. They include

- National Center for Atmospheric Research, Boulder, Colorado ($50 million annually), managed by University Consortium for Atmospheric Research;
- National Optical Astronomy Observatory (NOAO), three sites in Arizona, New Mexico, and Chile ($29 million a year), managed by Associated Universities for Research in Astronomy;
- National Radio Astronomy Observatory, Green Bank, West Virginia, and New Mexico ($27 million a year), managed by Associated Universities, Inc.;
- academic fleet, stationed at a number of universities ($50 million a year), their use managed by the University National Oceanographic Laboratory System, an association of institutions operating the ships for NSF and representatives of the academic oceanographic research community;
- Ocean Drilling Program ($36 million a year), managed by Joint Oceanographic Institutions, Inc.;
- Global Seismic Network and a portable seismic array for fine-detail local studies of the earth's crust ($7 million a year for operations), managed by Incorporated Research Institutions for Seismology; and
- Laser Interferometer Gravitational Wave Observatory (LIGO), Louisiana and Washington ($212 million to construct and an estimated $12 million a year to operate), to be built and managed by a collaboration between the California Institute of Technology and the Massachusetts Institute of Technology and secondary involvement by several other universities with interests in gravitational research.

APPENDIX B

A distinctive feature of the awards for these facilities is that they are not solicited competitively. The awards are based on the assumption that the facilities are national resources for the use of the entire research community, developed in conjunction with and managed by that community. Competition occurs when individual researchers submit proposals to use the facilities, which are evaluated by a selection committee including outside experts that is administered by the facility manager.

The projects must apply to renew their awards every three to five years. The renewal proposals are subject to the merit review process, just as standard investigator-initiated proposals are, usually involving mail reviews and site visits for input into the decision on whether or not to continue the activity. Proposals for upgrading or expanding facilities, such as NOAO's 8-meter GEMINI telescope, are handled as a separate award and reviewed separately, but still not using open competition.

Unique National Centers and Facilities Run by a University

Over the years the sharpest controversies over award decisions stem from these cases. The EERC and the National High Magnetic Field Laboratory (NHMFL) cases have been mentioned above. More recent cases include the National Nanofabrication Users Facility. As noted, the controversy surrounding the NHMFL stemmed from NSF's decision to make the award to the proposal ranked second by outside peer reviewers. The competing proposal ranked higher by peer reviewers came from the university that had had the most advanced high magnetic field facility in the world for many years and had been supported by NSF since 1972.

In the NHMFL and other single-facility cases, NSF faces the inherently difficult situation of choosing between a proposal from a long-established program or facility and a highly promising proposal from a place that has not had such a program or facility. These cases also place a high premium on procedural fairness, although they may reduce NSF's flexibility in reaching its program goals. For example, NSF opened itself to criticism in the EERC case for being inconsistent in applying criteria and procedures to competing proposers and thus appearing to prejudge the decision before the review process was complete.[1] Clear procedural fairness and consistency is also important because it encourages competition, especially in cases where there is an existing facility.

The Supercomputer Center facility awards are an intermediate case. They were highly desirable for the universities that received them. NSF's task of choosing among proposers who varied in their background experience and approaches while meeting several program goals was simplified because it was able to choose four winners, not one.

NSF supports a number of other expensive facilities located at and operated by universities, including the Cornell Electron Storage Ring, one of five national high-energy physics accelerators, and two of the seven accelerators available nationally for nuclear physics (the other accelerators are supported by the Department of Energy [DOE]). NSF also supports several university-based synchrotron light sources for materials research, research institutes for pure and applied mathematics and for theoretical physics, and the National Astronomy and Ionosphere Center operated by Cornell University in Arecibo, Puerto Rico.

International Projects

As research programs and supporting facilities become more expensive, international cooperation and financial participation become more desirable (and, with computer networks, more feasible). The Ocean Drilling Program (ODP), for example, has long had

---

[1] As a result of this incident, documented in a General Accounting Office (GAO, 1987) report, NSF revised the guidelines in its *Proposal and Award Manual* to require more uniformity in reviews of competing proposals and completion of the review process before an award decision is recommended.

substantial foreign participation. Six international partners (France, the Federal Republic of Germany, Japan, the United Kingdom, the European Science Foundation representing 12 smaller countries, and a consortium representing Canada and Australia) contribute $2.75 million a year each toward drilling costs and associated laboratories, core repositories, data banks, and engineering development activities. Each country separately funds the costs of the substantive research conducted by its scientists.

International participation creates a more complicated review process, because each member nation has to be consulted about the proposal and any changes in it. Such coordination issues were among the reasons that NSF decided to go ahead with the construction of two LIGO sites within the United States rather than wait to see if other countries would participate in an international effort involving additional sites around the world.

Another international project is the new 8-meter GEMINI telescope facility planned for Hawaii and Chile. Congress mandated the 50 percent foreign financial participation in the $178 million project that was only recently secured. Arrangements will now have to be worked out for joint reviews by the countries supporting the project.

BUDGET BREAKDOWN

Centers

Centers constituted 3 percent of the R&RA budget in FY 1982, 7 percent in FY 1990, and 8 percent ($145 million) in FY 1993. They would form a larger share of the budget if the initial plans for the ERC and STC programs had not been scaled back (NSF originally projected that they would be no more than 10 percent of the R&RA budget if it were doubled in five years, as proposed in FY 1987).

Facilities

Facilities are also growing as a share of the NSF budget, from 15 percent in FY 1982, to 18 percent in FY 1990, and 22 percent ($404 million) in FY 1993. More recently, this category has grown with the start of several major capital construction projects (e.g., LIGO [$212 million in NSF funding], the GEMINI telescopes [$79 million], and NHMFL [$60 million]).[2]

Disciplinary Research

NSF considers the majority (more than 70 percent) of the R&RA budget to be "disciplinary research." Most of the disciplinary research consists of individual research support, but it does include a few major awards. Among these are several large group projects in physics using DOE-funded national particle accelerator facilities, a long-term panel survey of income dynamics, and the ODP.

PROGRAM BALANCE

The overall balance among funding mechanisms varies by directorate (see Table B-1). Centers form a major part of the NSF engineering program (23 percent). Facilities are concentrated in the geosciences (35 percent), physical sciences (35 percent), and especially computer sciences and engineering (43 percent).

Most major awards are planned to complement traditional small research projects, for example, by providing them with access to expensive and/or unique facilities or facilitating the formation of a

---

[2] These are NSF obligations; the figures do not include matching funding from other sources. The state of Florida is contributing an additional $58 million to the NHMFL; other countries are contributing an additional $79 million to the GEMINI project.

TABLE B-1: Balance (percent) Among NSF Research Funding Mechanisms by Directorate, FY 1992

	Research Directorate						
Mechanism	BIO	CISE	ENG	GEO	MPS	SBE	All
Disciplinary research	96.5	53.0	75.9	64.0	64.8	99.3	71.1
Facilities	0.0	42.9	1.2	34.8	27.7	0.0	22.0
Centers	3.5	4.2	22.9	1.2	7.6	0.7	7.0

NOTES: Some NSB-approved major awards are classified as disciplinary research. BBS—Biological Sciences; CISE—Computer and Information Science and Engineering; ENG—Engineering; GEO—Geosciences; MPS—Mathematical and Physical Sciences; SBE—Social, Behavioral and Economic Sciences.

SOURCE: NSF Executive Information System.

critical mass of researchers with skills needed to address an important research problem.

Astronomy

Astronomy is an example of a field in which most individual researchers are dependent on large-scale facilities to conduct their work. Two-thirds of the funding for astronomy goes to the national observatories, and part of the remaining one-third goes to university-based telescopes. Many individual investigator grants support the researchers while they are using the national facilities.

Materials Science

A little more than half the funding for materials science goes to individual investigators and small groups. The rest goes to MRLs and the national facilities (synchrotrons, high magnetic field laboratory, etc.) and to a fast-growing activity, collaborative materials research groups. Individual grants help underwrite the costs of researchers using the national facilities.

Geosciences

In the atmospheric sciences, 52 percent of the funding goes to research project grants; the rest goes to the National Center for Atmospheric Research and other facilities. The ocean sciences are also very dependent on facilities; 52 percent goes to research project grants and the rest to the ODP, the academic fleet, and other facilities.

C Awards Approved by National Science Board, FY 1986-1992

NSB Meeting	Project Title	Institution	Award Size ($ millions)
11/92	Experimental Particle Physics Program	Columbia University	$ 11.70
	Science and Technology Center (STC) for Superconductivity	University of Illinois, Urbana	9.00
	National Nanofabrication Users Facility	(project development plan)	0.00
		Total	$ 20.70
10/92	Urban Systemic Initiatives in Science, Mathematics, and Technology Education	(program)	0.00
	Polar Ice Coring and Logistic Support Activities	(request for proposals)	26.00
	Support of Operation of Nuclear Structure Laboratory and Research	State University of New York (SUNY)—Stony Brook	9.24
		Total	35.24
8/92	National Nanofabrication Users Facility	Cornell University	2.00
	Management/Operation of the Ocean Drilling Program	Joint Oceanographic Institutions (JOI)	309.50
	Ocean Science Accelerator Mass Spectrometry Facility	Woods Hole Oceanographic Institution	6.43
	Synchrotron Radiation Center	University of Wisconsin	11.55
	Superconducting Cyclotron Laboratory	Michigan State University	32.90
	Interactive Mathematics Project Phase II	San Francisco State University	9.06
	1993 National Survey of College Graduates	U.S. Census Bureau	9.60
		Total	381.01
6/92	National Center for Atmospheric Research	University Consortium for Atmospheric Research (UCAR)	107.00
	Materials Research Laboratory	University of California, Santa Barbara	11.05
	Materials Research Laboratory	Cornell University	15.42
	Materials Research Laboratory	Northwestern University	8.1
	Materials Research Laboratory	University of Pennsylvania	11.4
	Industry/University Cooperative Research Center for Simulation and Optimization of Mechanical Systems	University of Iowa	17.03
		Total	170.00

5/92	STC for Molecular Biotechnology	University of Washington	14.29
	California Mathematics and Science Initiative	California Department of Education	10.00
	Georgia Statewide Systemic Middle School Program	University of Georgia Research Foundation	10.00
	Partnerships for Reform Initiatives in Science and Mathematics	Kentucky Department of Education	9.67
	Maine: A Community of Discovery Partnerships Advancing Learning of Science and Math	Maine Department of Education	10.00
	State Systemic Initiative (SSI) for Mathematics and Science	Massachusetts Department of Education	9.74
	SSI in Science and Mathematics Education	Michigan Department of Education	9.99
	SSI in Science and Mathematics Education	University of New Mexico	9.94
	Texas Science and Mathematics Renaissance	University of Puerto Rico	10.00
	SSI in Science, Mathematics, and Technology	Texas Education Agency	10.00
	Engineering Design Research Center	Vermont Department of Education	9.62
	Engineering Research Center (ERC) for Advanced Technology for Large Structural Systems	Carnegie Mellon University	13.40
		Lehigh University	12.50
	ERC for Compound Semiconductor Microelectronics	University of Illinois	13.57
	ERC for Net Shape Manufacturing	Ohio State University	11.30
	ERC for Advanced Combustion	Brigham Young University/University of Utah	11.71
		Total	185.43
3/92	Board on International Comparative Studies in Education	National Academy of Sciences	8.19
	Increased Support for the Presidential Awards for Excellence in Science and Math Teaching	National Science Teachers Association (NSTA)	12.32
	The Teachers' Academy of Science in Chicago	Teachers Academy—Mathematics/Science, Chicago	6.50
	Research Support	International Institute for Applied Systems Analysis (IIASA)	6.31
		Total	33.32
2/92	STC for High Performance Polymeric Adhesives and Composites	Virginia Polytechnic Institute/State University	7.49
11/91	STC for Microbial Ecology	Michigan State University	10.02
	STC for Advanced Cement-Based Materials	Northwestern University	10.25
	STC for Research on Parallel Computation	Rice University	23.50

NSB Meeting	Project Title	Institution	Award Size ($ millions)
11/91 (cont.)	STC for Quantized Electronic Structures	University of California, Santa Barbara	12.25
	STC for Analysis and Prediction of Storms	University of Oklahoma	9.00
	STC for Photoinduced Charge Transfer	University of Rochester	9.63
	STC for Discrete Mathematics and Theoretical Computer Science	Rutgers University	10.00
	STC for Particle Astrophysics	University of California, Berkeley	12.96
	STC for Superconductivity	University of Illinois, Urbana	13.50
	Gateway Engineering Education Coalition	Drexel University	15.00
	Southeastern University and College Coalition for Engineering Education	North Carolina State University	15.00
	Total		141.11
10/91	ERC for Interfacial Engineering	University of Minnesota	14.45
	ERC for Advanced Electronics Materials Processing	North Carolina State University/Duke University/University of North Carolina—Chapel Hill	14.80
	ERC for Offshore Technology	Texas A&M/University of Texas, Austin	14.20
	ERC for Plasma-Aided Manufacturing	University of Wisconsin, Madison/University of Minnesota, Minneapolis	14.55
	Total		58.00
8/91	Francis Bitter National Magnet Laboratory	Massachusetts Institute of Technology	23.00
	Mathematical Sciences Research Institute	Mathematical Sciences Research Institute	15.84
	Institute for Mathematics and Its Applications	University of Minnesota	9.60
	Square One TV: Two Additional Seasons	Children's Television Workshop (CTW)	2.50
	Total		50.94
5/91	The Panel Study of Income Dynamics	University of Michigan	$ 13.06
	SSI - Connecticut Project CONSTRUCT	Connecticut State Department of Education	7.87
	SSI - Delaware Project 21	Delaware Department Public Instruction	4.95
	SSI - Florida Comprehensive Plan for Mathematics/Science/Computer Education	Florida Department of Education	7.91

	SSI - Louisiana Systemic Initiatives Program	Louisiana Board of Regents	10.00
	SSI - Montana Challenge 2001/Mathematics (SIMM)	Montana Council of Teachers of Mathematics	9.94
	SSI - Nebraska Statewide Initiative	University of Nebraska, Lincoln	4.47
	SSI - North Carolina Science/Mathematics Alliance	North Carolina Science/Mathematics Alliance	4.86
	SSI - Ohio Mathematics/Science Discovery Project	Ohio Board of Regents	10.00
	SSI - Rhode Island SSI for Science and Math	Rhode Island Dept. Elementary/Secondary Education	9.35
	SSI - South Dakota SSI	South Dakota Department of Education	7.52
	National Center for Earthquake Engineering Research, Phase II	SUNY—Buffalo	21.00
	Cooperative Research in Integrated Circuit Technology	Semiconductor Research Corporation	12.50
	Development of Casting/Polishing - 8-Meter Mirrors	University of Arizona	1.62
	Program Development Plan/Phase I Award: GEMINI 8-Meter Telescopes	Associated Universities for Research in Astronomy (AURA)	20.00
	Support for US Participation in International Congress of Scientific Unions	National Academy of Sciences	3.25
		Total	151.30
3/91	ERC for Cardiovascular Technology	Duke University	13.95
	ERC for Telecommunications Research	Columbia University	14.60
	ERC for Bioprocess Engineering	Massachusetts Institute of Technology	11.45
	ERC for Systems Research	University of Maryland/Harvard University	13.85
	ERC for Intelligent Manufacturing Systems	Purdue University	16.86
	Earth's Dynamics and Structure	Incorporated Research Institutions for Seismology (IRIS)	75.00
	Materials Research Laboratory	Massachusetts Institute of Technology	11.66
	Materials Research Laboratory	Stanford University	9.42
	Research Support	IIASA	2.00
		Total	168.79
2/91	U.S. Science Support for Ocean Drilling	JOI	14.00
	Presidential Awards for Science/Mathematics Teaching	NSTA	4.60
		Total	18.60

NSB Meeting	Project Title	Institution	Award Size ($ millions)
11/90	National Radio Astronomy Observatory	Associated Universities, Inc.	69.59
	Extension of NSFNET Backbone for T-3	MERIT, Inc.	8.00
	Mathematical Sciences Education Board	National Academy of Sciences	10.00
	Total		87.59
10/90	Laser Interferometer Gravitational Wave Observatory (LIGO)	California Institute of Technology	30.60
	Cyclotron Studies of Processes at Intermediate Energy	Indiana University	12.36
	Research in Atomic and Molecular Physics	University of Colorado	
	Total		42.96
8/90	Extension and Utilization of Seven Seasons of Science Programming for Children	CTW 3-2-1 Contact	5.51
	An Elementary Mathematics Curriculum	TERC	6.17
	Engineering Education Coalition	Cornell University	15.34
	Engineering Education Coalition	Howard University	15.00
	National High Magnetic Field Laboratory	Florida State University	60.00
	STC for Biological Timing	University of Virginia	5.67
	STC for Engineering Plants for Resistance Against Pathogens	University of California, Davis	5.44
	STC for Light Microscope Imaging and Biotechnology	Carnegie Mellon University	3.41
	STC for Research in Cognitive Sciences	University of Pennsylvania	5.54
	STC for Magnetic Resonance Technology for Basic Biological Research	University of Illinois	5.50
	STC for Advanced Liquid Crystalline Optical Materials	Kent State University	5.45
	STC for Synthesis, Growth, and Analysis of Electrical Materials	University of Texas, Austin	5.00
	STC for Clouds, Chemistry, and Climate	University of Chicago	4.13
	STC for Astrophysical Research in Antarctic	University of Chicago	1.73
	STC for High Pressure Research	SUNY-Stony Brook	8.37
	STC: Southern California Earthquake Center	University of Southern California	
	STC for Computation and Visualization of Geometric		

	Structures		
	STC for Ultrafast Optical Science	University of Minnesota	6.52
	College and University Innovation Research	University of Michigan	5.52
		(pilot program)	0.00
	Evaluation Services for Graduate Research Fellowships	(request for proposals)	0.00
		Total	**179.97**
5/90	Operation of the National Astronomy and Ionosphere Center, Arecibo	Cornell University	43.54
	Second Arecibo Upgrading Program Phase I and II	Cornell University	22.80
	LIGO	California Institute of Technology	212.00
		Total	**278.34**
3/90	ERC for Optoelectronic Computing Systems	Colorado University	15.60
	Graduate Engineering Program for Women, Minorities, and Disabled	(program)	0.00
	NSF Alliances for Minority Participation	(program)	0.00
	Statewide Systemic Initiatives	(program)	0.00
		Total	**15.60**
2/90	Nuclear Research with Electrons, Photons, Anti-Photons	Illinois University	7.41
10/89	Biophysical Studies on Macromolecular Assemblies	Johns Hopkins University	$0.25
	Design of Green Bank Telescope	National Radio Astronomy Observatory	4.41
	Institute for Theoretical Physics	University California, Santa Barbara	14.30
	ERC for Data Storage Systems	Carnegie Mellon University	14.60
	ERC for Geometrically-Complex Field Problems	Mississippi State University	12.20
	Program for Arctic Social Science	(program)	0.00
		Total	**52.96**
8/89	Research on Gigabit Networks	CNRI	15.00
	8-Meter Telescope Mirrors at Steward Observatory	University of Arizona	3.78
	National Superconducting Cyclotron Laboratory	Michigan State University	33.51
	Children's TV Series About Mathematics	CTW	6.60
	Research Training Group Program	(program)	0.00
		Total	**58.89**

NSB Meeting	Project Title	Institution	Award Size ($ millions)
6/89	Increased Funding for NSFNET Backbone Network	Merit, Inc.	6.00
5/89	San Diego Supercomputer Center	General Atomics	88.50
	National Center for Supercomputing Applications	University of Illinois	84.70
	Cornell Theory Center/National Supercomputer Facility	Cornell University	79.10
	3-2-1 Contact	CTW	5.51
		Total	**346.41**
3/89	ERC for Advanced Combustion Engineering Research	Brigham Young University	12.00
	ERC for Engineering Design	Carnegie Mellon University	17.00
	ERC for Compound Semiconductor Microelectronics	University of Illinois, Urbana	15.00
	ERC for Advanced Technology for Large Structural Systems	Lehigh University	15.00
	ERC for Net Shape Manufacturing	Ohio State University	14.00
		Total	**73.00**
2/89	Management, Operation, and Maintenance of the National Optical Astronomy Observatory	AURA	166.45
12/88	STC for Photoinduced Charge Transfer	Rochester University	8.85
	STC for Discrete Mathematics and Theoretical Computer Science	Rutgers University	9.82
	STC for Particle Astrophysics	University of California, Berkeley	10.62
	STC for High Performance Polymeric Adhesives and Composites	Virginia Polytechnic Institute/State University	7.85
	STC for Advanced Cement/Based Materials	Northwestern University	9.75
	STC for High Temperature Superconductivity	Illinois University	24.25
	STC for Quantized Electronic Structures	University of California, Santa Barbara	11.70
	STC for Development for Research on Parallel Composites	Rice University	22.90
	STC for Integrated Protein and Nucleic Acid Biotechnology	California Institute of Technology	17.05

	STC in Microbial Ecology	Michigan State University	7.10
	STC for Analysis and Prediction of Storms	University of Oklahoma	4.90
	Replacement of the Research Vessel CONRAD	Columbia University	11.38
	Research at the Sondrestrom Radar Facility	SRI International	7.77
	National High Magnetic Field Laboratory	(program)	0.00
	Icebreaking Capability for Use in U.S. Antarctic Program (USAP)	ITT Antarctic Services	165.00
		Total	318.94
10/88	Ocean Science Accelerator Mass Special Facility	Woods Hole Oceanographic Institution	5.40
5/88	Biological Research Center for Insect Science	University of Arizona	1.72
	Biological Research Center for Plant Developmental Biology	University of California, Berkeley	2.00
	Biological Research Center for Biophysical Studies	Johns Hopkins University	2.20
	Massive Memory Machine Project	Princeton University	1.10
	Laser Gravitational Wave Detector	California Institute of Technology	10.60
	Development Plan for Research Interagency Backbone	(project development plan)	0.00
	Specialized Support for USAP	(request for proposals)	0.00
	Graduate Fellowships for Women in Engineering Program	(program)	0.00
		Total	17.62
3/88	ERC for Telecommunications Research	Columbia University	20.90
	ERC for Systems Research Center	Harvard University	21.30
	ERC for Biotechnology Process	Massachusetts Institute of Technology	20.00
	ERC for Intelligent Manufacturing	Purdue University	17.67
		Total	79.87
2/88	Cornell High Energy Synchrotron Source	Cornell University	6.40
	Science Support for the Ocean Drilling Program	JOI	14.20
	Polar Ice Coring and Logistic Support	(request for proposals)	0.00
		Total	20.60

NSB Meeting	Project Title	Institution	Award Size ($ millions)
10/87	NSFNET Backbone Network	MERIT, Inc.	14.00
	Icebreaking Capability for Use in the USAP	UCAR	107.00
	National Center for Atmospheric Research		297.00
	NSF Award to Semiconductor Research Corporation for SEMATECH	Semiconductor Research Corporation	3.00
		Total	421.00
8/87	Pittsburgh Supercomputing Center	Mellon-Pitt-Carnegie Corporation	30.00
	National Center for Atmospheric Research	UCAR	61.07
	A 450 MeV Cascade Microtron	University of Illinois, Urbana	23.00
	Square-One TV Mathematics Program	CTW	9.00
	Minority Research Center of Excellence (MRCE)	Howard University	5.00
	MRCE: Center for Cellular and Molecular Biology	Meharry Medical College	5.00
	Research Career Innovation and Development	(program)	0.00
	Science and Technology Research Centers	(program)	0.00
		Total	133.07
5/87	Memorandum of Understanding with Semiconductor Research Corporation	Semiconductor Research Corporation	10.00
	John von Neumann Center for Advanced Supercomputing	Consortium for Scientific Computing	38.50
	Management and Operation of the National Science Research Network (NSFNET)	(project development plan)	0.00
	Postdoctoral Research Fellowships in History and Philosophy of Science	(program)	0.00
	Young Scholars Program	(program)	0.00
	Undergraduate Faculty Enhancement Program	(program	0.00
		Total	48.50
3/87	ERC for Hazardous Substance Control	University of California, Los Angeles	18.00
	ERC for Optoelectronic Computing	University of Colorado	14.50
	ERC for Emerging Cardiovascular Technology	Duke University	14.00
	Materials Research Laboratory	Stanford University	2.95

	Materials Research Laboratory	Brown University	1.80
	Materials Research Laboratory	Harvard University	5.78
	Materials Research Laboratory	University of Illinois, Urbana	9.97
	Large Scale Nonlinear Systems Engineering Program	(program)	0.00
		Total	**67.00**
2/87	Cornell Nanofabrication Facility	Cornell University	10.00
	New Lease Arrangements for POLAR DUKE		20.60
	Computational Engineering Program	(program)	0.00
	Neuroengineering Program	(program)	0.00
	Operating Plan for MRCE Program	(operating plan)	0.00
		Total	**30.60**
11/86	Deep Observation and Sampling of the Earth's Continental Crust	DOSECC, Inc.	25.80
8/86	National Center for Supercomputing Applications	University of Illinois, Urbana	31.60
	San Diego Supercomputer Center	GA Technologies	44.00
	Advanced Supercomputer Center (John von Neumann Center for Scientific Computing)	Consortium for Science Computing	11.50
	IRIS	IRIS	26.10
	Seismic Reflection Profiling	Cornell University	6.42
	National Center for Atmospheric Research	UCAR	55.39
	Inner Shelf Transfer and Recycling	University of Alaska	5.00
	Experimental Program to Stimulate Competitive Research (EPSCoR)		
	EPSCoR Implementation Award	University of Alabama	3.00
	EPSCoR Implementation Award	University of Kentucky	3.00
	EPSCoR Implementation Award	University of Nevada	3.00
	EPSCoR Implementation Award	North Dakota Higher Education	3.00
	EPSCoR Implementation Award	Oklahoma State University	3.00
	EPSCoR Implementation Award	University of Puerto Rico	3.00
	EPSCoR Implementation Award	University of Vermont	2.50
	EPSCoR Implementation Award	University of Wyoming	3.00
	Engineering Research Center	SUNY—Buffalo	25.00
		Total	**228.51**

NSB Meeting	Project Title	Institution	Award Size ($ millions)
5/86	Panel Study of Income Dynamics	University of Michigan	10.00
	Synchrotron Radiation Center	University of Wisconsin, Madison	8.75
		Total	**18.75**
3/86	ERC for Engineering Design	Carnegie Mellon University	14.90
	ERC for Compound Semiconductor Microelectronics	University of Illinois, Urbana	11.60
	ERC for Advanced Technology for Large Structural Systems	Lehigh University	10.40
	ERC for Net Shape Manufacturing	Ohio State University	9.70
	ERC for Advanced Combustion Engineering Research	Brigham Young University	9.70
	Materials Research Laboratory	Massachusetts Institute of Technology	14.30
	Materials Research Laboratory	University of Chicago	3.60
	Materials Research Laboratory	Cornell University	14.40
	Materials Research Laboratory	Northwestern University	7.02
	Materials Research Laboratory	University of Pennsylvania	11.37
	Physics Laboratory	University of Chicago	9.90
		Total	**116.89**
1/86	Astronomy Studies with Submillimeter Observatory	California Institute of Technology	4.63
	The University NAVSTAR Consortium (UNAVCO)	University of Colorado	2.30
	Interactions of Muons, Kaons, Antiprotons and Sigma Hyperons	William and Mary College	1.68
		Total	**14.61**
11/85	FY 86-90 Management, Operation, and Maintenance of National Astronomy and Ionosphere Observatory	Cornell University	43.51
	Deep Observation and Sampling of the Earth's Continental Crust	DOSECC, Inc.	3.02
	IRIS - 1986 Program	IRIS	2.50
	Mathematical Science Research Institute	Mathematical Science Research Institute	13.00
	Institute for Mathematics and its Applications	University of Minnesota	6.50
	Theory of Elementary Particles	University of California, Berkeley	2.47

Nuclear Research with a Tandem Accelerator	University of Pennsylvania	3.06
Theoretical Physics	Princeton University	3.06
Operation of the Nuclear Structure Research Laboratory	Rochester University	3.47
Support of a Cosmic Ray Observatory for Ultra High Energy Processes	University of Utah	3.60
Advanced Scientific Computing Resources	Digital Productions	1.00
Assessment of Initiatives to Address Problems and Opportunities in Science Education	(request for proposals)	2.00
Interacademy Science Exchanges with the USSR	National Academy of Sciences	1.90
	Total	89.09

D Major Award Criteria from Recent Solicitation Announcements

<u>Engineering Research Centers (ERCs)</u> (FY 1990)

According to the Program Announcement for ERCs issued in 1988 for FY 1990 awards, the criteria used in the evaluation of proposals were based on the guidelines for merit review in the National Science Foundation (NSF, 1992d) document *Grants for Research and Education in Science and Engineering* and on the key features of an ERC. They were

1. research merit and potential impact on U.S. competitiveness;
2. strength and impact of educational programs;
3. industrial/other user participation and knowledge/technology transfer;
4. leadership and performance competence;
5. institutional environment and support;
6. effect on the infrastructure of engineering.

<u>NSF Science and Technology Centers (STCs)</u> (FY 1990)

STC proposals went through a three-stage review process. In phase one, mail and panel reviewers were asked to evaluate proposals based on the selection criteria below. In phase two, the most promising proposals underwent a site visit. In phase three, a special review panel was convened to recommend awards by considering "the relative merit of the proposals using the criteria listed below, the balance of awards among scientific fields, and the combined ability

APPENDIX D

of the centers to meet the objectives of the STC program, as well as to enhance the Nation's economic competitiveness":

1a. intrinsic merit of the research;
1b. research performance competence;
2. effect of the center on the infrastructure of science and engineering;
3. rationale for the center;
4. utility or relevance of the research;
5. institutional support and management plan.

Earthquake Engineering Research Center (FY 1986)

In addition to the basic four criteria described in *Grants for Research and Education in Science and Engineering* (NRC, 1992d), the following criteria were "taken into consideration in rating the proposals":

- the relevance of the center to the NSF role in earthquake hazard mitigation;
- relevance of the selected research center problem area to earthquake hazard mitigation;
- demonstrated capability to manage, direct, and focus research center activities to establish a coordinated and directed effort in the problem area;
- detailed statements of objectives, goals, and mission of the research center and the methodology for achievement;
- management plan and methodology to allow center activities to be directed, coordinated, and focused;
- plan and methodology for incorporating earthquake researchers from other institutions, industry, and government into the research center's activities;

- plan and methodology for integrating the education of engineers into the research center to provide highly trained professionals in earthquake engineering; and
- plan and methodology for effective and accelerated technology transfer of research results to the end users and the subsequent solution of the relevant problem area.

National High Magnetic Field Laboratory (NHMFL) (FY 1990)

According to the solicitation announcement, proposals for the NHMFL were to be evaluated by using a two-stage review process. In stage one, mail and panel reviews were used to identify the most meritorious proposals. In stage two, the most meritorious proposals were visited by a site visit team of experts. In this case, the top proposals were also reviewed by the NSF Materials Research Advisory Committee. The following criteria were to be used in the selection process:

1a. intrinsic merit of the research;
1b. research performance competence;
2. effect of NHMFL on the infrastructure of science and engineering:

    a. the form, appropriateness, effectiveness, and strength of scientific and technical connections and exchanges with other sections and groups;
    b. the quality and appropriateness of the educational and training components;

3. utility or relevance of the research;
4. management plan;
5. institutional and other sector support.

National Center for Geographic Information and Analysis (FY 1988)

The four general criteria described in *Grants for Research and Education in Science and Engineering* (NSF, 1992d) were employed. along with and the degree to which each proposal contained the following:

- a clear identification of basic research problems in geographic analysis and geographic information systems that will be the focus of the center's activities. Such problems must be of common concern to scholars in a number of academic disciplines;
- specific plans for productive, multidisciplinary cooperation among faculty, students, and GIS [geographic information systems] practitioners on topics of mutual interest and concern;
- programs that will help alleviate the serious shortage of personnel trained in geographic information systems and geographic analysis in the public, private, and academic sectors;
- plans for acting as a clearinghouse and conduit for information regarding the existence, characteristics, and availability of geographic data bases, domestically and internationally;
- measures designed to maintain and enhance the international competitiveness of the United States with respect to geographic analysis and geographic information systems;
- significant commitments of institutional funds and a plan for obtaining support from external sources in the forms of funds, equipment, and personnel that ensure the involvement of practitioners in the center's research and instructional programs; and
- a management plan for the center that assures broad and continued participation in center oversight by scholars and practitioners from throughout the nation's GIS community.

Materials Research Laboratories (MRL) (FY 1992)

Criteria for evaluation of MRL proposals were drawn from the guidelines for merit review in *Grants for Research and Education in Science and Engineering* (NSF, 1992d) and the following key features of an MRL:

1. Research thrust areas

    - Intrinsic merit of the research
    - Research performance competence
    - Degree of interconnection

2. The MRL as a whole

    - Institutional setting and rationale for the MRL
    - Central facilities
    - Seed funding
    - Effect of the MRL on the infrastructure of science and engineering
    - Institutional support and management plan

E   *The Ten Case Studies*

1. IRIS (Incorporated Research Institutions for Seismology)
2. Engineering Research Centers
3. National Nanofabrication Facility
4. Earthquake Engineering Research Center
5. Ocean Drilling Program
6. National High Magnetic Field Laboratory
7. GEMINI 8-Meter Telescopes
8. Science and Technology Centers
9. Supercomputer Centers
10. LIGO (Laser Interferometer Gravitational Wave Observatory)

References

Bloch, Erich
    1986   *Annual Report on the Foundation's Use of Peer Review.* NSB-86-45. February 25.

England, J. Merton
    1982   *A Patron for Pure Science: The National Science Foundation's Formative Years, 1945-57.* NSF 82-24. Washington, D.C.: National Science Foundation.

General Accounting Office
    1982   *NSF's Award of Two Research Vessels Met Requirements.* GAO/PAD-82-14. Washington, D.C.: U.S. Government Printing Office.
    1987   *National Science Foundation: Problems Found in Decision Process for Awarding Earthquake Center.* GAO/RCED-87-146. Washington D.C.: U.S. Government Printing Office.

Massey, Walter E.
    1991   *Annual Report on the NSF Proposal Review System.* NSB-9-68. April 24.
    1992   *FY 1991 Report on the NSF Merit Review System.* NSB-92-42. March.

NAE (National Academy of Engineering)
    1983   *Strengthening Engineering in the National Science Foundation: A View from the President of the National Academy of Engineering.* Washington D.C.: National Academy of Engineering.
    1984   *Guidelines for Engineering Research Centers: A Report for the National Science Foundation by the National Academy of Engineering.* Washington, D.C.: National Academy of Engineering.
    1985   *New Directions for Engineering in the National Science Foundation: A Report to the National Science Foundation from*

REFERENCES

    National Academy of Engineering. Washington D.C.: National Academy of Engineering.

NAS (National Academy of Sciences)
- 1987 *Science and Technology Centers: Principles and Guidelines.* Washington, D.C.: National Academy of Sciences.
- 1989 Federal Science and Technology Budget Priorities: New Perspectives and Procedures. Ch. V in *National Issues in Science and Technology.* Washington, D.C.: National Academy Press.
- 1993 *Science, Technology, and the Federal Government: National Goals for a New Era.* Committee on Science, Engineering, and Public Policy, National Academies of Science and Engineering and Institute of Medicine. Washington, D.C.: National Academy Press.

NRC (National Research Council)
- 1991 *The Decade of Discovery in Astronomy and Astrophysics.* Astronomy and Astrophysics Survey Committee, Commission on Physical Sciences, Mathematics, and Applications. Washington, D.C.: National Academy Press.

NSB (National Science Board)
- 1967 Criteria for the Support of Research by the National Science Foundation. NSB-67-133. Approved by the National Science Board at its 112th Meeting, May 18-19.
- 1974 Criteria for the Selection of Research Projects by the National Science Foundation. NSB-74-300. As approved by the National Science Board at its 167th Meeting, October 17-18.
- 1979 "Big Science" Policies and Procedures Statement. NSB-79-65. Adopted by the National Science Board at its 203rd Meeting, January 18-19.
- 1983a *The Engineering Mission of NSF over the Next Decade.* Vol. I, Issues and Analysis; Vol. II, Background Material. NSB-83-127. June.
- 1983b Statement on the Engineering Mission of the NSF over the Next Decade. NSB-83-250. Adopted by the National Science Board at its 246th Meeting, August 18-19.
- 1983c Agenda Item 7, Minutes of the 247th Meeting of the National Science Board, September.
- 1986 Report of the National Science Board Committee on Leveraging National Science Foundation Funds. NSB-86-95. May 27.
- 1988a Report of the National Science Board Committee on Centers and Individual Investigator Awards. NSB-88-35. February.

1988b Minutes of the Executive Committee of the National Science Board. NSB/EC-88-9. August 18.
1989 *Leveraging NSF Funds: Update of the 1986 NSB Committee Report.* NSB-89-167. September 29.
1991a *Background Material for Long-Range Planning: 1993-1997.* NSB-91-105. June 20.
1991b Minutes of the 298th Meeting of the National Science Board, Open Session. NSB-91-25. February 14.
1992 Management and Operations of the Ocean Drilling Program, Memorandum to the Members of the National Science Board. NSB-92-131. August 3.

NSF (National Science Foundation)
1967 *Annual Report of the National Science Foundation for Fiscal Year 1967.*
1973 MRL Program Policy Statement. February.
1975 *Peer Review and Proposal Evaluation Staff Study.* June.
1977 Processing Recommendations Requiring National Science Board Approval. NSF Circular No. 107, Revision No. 2. April 21.
1982 Report of the NSF Panel on Large Scale Computing in Science and Engineering, chaired by Peter D. Lax. December 26.
1983 A National Computing Environment for Academic Research, prepared by the NSF Working Group on Computers for Research. July.
1984a Program Announcement: Engineering Research Centers (Centers for Cross-Disciplinary Research in Engineering), Fiscal Year 1985.
1984b Report of the Task Force on Bureaucracy and Control. December.
1986a Final Report of the NSF Advisory Committee on Merit Review, chaired by Norman Hackerman. NSF-86-93. September 25.
1986b NSF Director's Statement on the Report of the Advisory Committee on Merit Review. NSF-86-93. December 17.
1987 Program Announcement: Engineering Research Centers (Centers for Cross-Disciplinary Research in Engineering), Fiscal Year 1988.
1988a Report of the NSF Panel on Large Magnetic Fields, chaired by Frederick Seitz and Robert C. Richardson. July.
1988b Program Announcement: Engineering Research Centers (Centers for Cross-Disciplinary Research in Engineering).
1989 National High Magnetic Field Laboratory: Proposal Solicitation.
1990a Planning and Priority-Setting in the National Science Foundation: A Report to the Committee on Science, Space and Technology, U.S. House of Representatives. February 28. Pp. 350-369 in

REFERENCES

U.S. Congress, *Fiscal Year 1992 and 1993 National Science Foundation Authorization.* Hearings before the Science Subcommittee of the Committee on Science, Space and Technology, U.S. House of Representatives. 102d Congress, 1st session. February 20, March 12-13, 1991.

1990b Facilities Planning at the National Science Foundation: A Report to the Committee on Science, Space and Technology, U.S. House of Representatives. December 31. Pp. 341-349 in U.S. Congress, *Fiscal Year 1992 and 1993 National Science Foundation Authorization.* Hearings before the Science Subcommittee of the Committee on Science, Space and Technology, U.S. House of Representatives. 102d Congress, 1st session. February 20, March 12-13, 1991.

1990c *Report of the Merit Review Task Force.* NSF-90-113. August 23.

1990d Changes to NSF's Proposal Review System. Notice No. 109. July 17.

1991 *Justification of Estimates of Appropriations to the Congress, Fiscal Year 1993.*

1992a *Proposal and Award Manual.* NSF Manual No. 10. March.

1992b Materials Research Groups: Program Guideline. NSF 92-43. April 27.

1992c *Engineering Research Center: Program Solicitation.* Never released.

1992d *Grants for Research and Education in Science and Engineering: An Application Guide.* NSF-92-89.

1993a Design, Review, and Management Protocol. NSF Staff Memorandum O/D 93-02. January 5.

1993b Guidelines for Preparation of Items for NSB Review Action Items, Information Items, and Requests for Waivers. February.

1993c *Proposal and Award Manual.* NSF Manual No. 10. February.

Schwartz, Lyle H.
1987 Materials research laboratories: Reviewing the first twenty-five years. Pp. 35-48 in Peter A. Psaras and H. Dale Langford, eds., *Advancing Materials Research.* Washington, D.C.: National Academy Press.

Sproull, Robert L.
1987 Materials research laboratories: The early years. Pp. 25-34 in Peter A. Psaras and H. Dale Langford, eds., *Advancing Materials Research.* Washington, D.C.: National Academy Press.

U.S. Congress
  1992 *Hearings on Departments of Veterans Affairs and Housing and Urban Development, and Independent Agencies Appropriations for 1993.* Part 2. Subcommittee on VA, HUD, and Independent Agencies, House Committee on Appropriations. Washington, D.C.: U.S. Government Printing Office.

Windus, Margaret L.
  1984 *National Science Board Policy Activity over the Past 10 Years.* October. NSB-84-300. Washington, D.C.: National Science Board.